GENES
and the
ENVIRONMENT

GENES
and the
ENVIRONMENT

ROY H BURDON

UNIVERSITY OF STRATHCLYDE,
GLASGOW, SCOTLAND

UK	Taylor & Francis Ltd, 11 New Fetter Lane, London EC4P 4EE
USA	Taylor & Francis Inc., 325 Chestnut Street, 8th Floor, Philadelphia, PA 19106, USA

Taylor & Francis is an imprint of the Taylor & Francis Group

Every effort has been made to ensure that the advice and information in this book is true and accurate at the time of going to press. However, neither the publisher nor the authors can accept any legal responsibility or liability for any errors or omissions that may be made. In the case of drug administration, any medical procedure or the use of technical equipment mentioned within this book, you are strongly advised to consult the manufacturer's guidelines.

British Library Cataloguing in Publication Data
A catalogue record for this book is available from the British Library.

ISBN 0–7484–0827–4 (HB)
ISBN 0–7484–0826–6 (PB)
Library of Congress Cataloging in Publication data are available

Typeset by Graphicraft Limited, Hong Kong
Printed by T.J. International Ltd, Padstow, UK

Contents

① Gene Damage: Its Limitation, Repair and Consequences

(2)

Environmental Stress: Genomic Responses

The Environment, Genes, Biodiversity and Cancer 221

Preface

Our interactions with the environment are complex, but crucial to these are the precise molecular effects of environmental factors in relation to the DNA of genes. This is a subject treated scantily in most present-day general biochemistry, molecular biology, and even environmental textbooks. Such interactions are of wide-reaching significance for the mechanisms of adaptation, acclimation and evolution. It is also clear that they are very relevant to the development of important human degenerative diseases such as cancer, and to the processes that contribute to our ageing.

This book is primarily concerned with the molecular biology of microbial, plant and animal genes, and how their chemical structure and activity can be influenced by a range of currently relevant environmental adversities. It begins with a broad overview of the different types of DNA damage that can be inflicted by, for example, toxic synthetic chemicals, ultraviolet light, ionizing radiation, food components, and atmospheric pollutants such as those from fossil fuels, cigarette smoke and automobile exhaust. The mechanisms that exist in cells to minimize such damage and to repair the damaged genes are then illustrated. The short-term repercussions of inadequate damage limitation for human health and disease are highlighted as well as the consequences for long-term evolutionary patterns.

While organisms can limit specific chemical damage to genes, a later section of the book focuses on the mechanisms whereby a specific class of genes can be activated to yield special proteins that are used to protect vital cellular structures from environmental adversity. The molecular mechanisms involved in the activation of many such protective 'stress genes' are now well understood and will serve as useful paradigms for future explorations of more complex interactions between genes and the environment.

Although growing importance is now attached by governments and health authorities to the increasing threats arising from our environment, the consideration of environmental issues from a molecular biological standpoint is especially timely. The following recent developments contribute to this view.

- Precise knowledge of the layout of gene sequences that comprise the human genome is now nearing completion. We already have such information in relation to a number of key microbes, and understanding of the nucleotide sequences of representative plant genomes is well advanced.

- There has been a very significant growth in our appreciation of *free radicals* and how they can damage cellular components, including DNA. This provides new molecular insights into the genetic hazards of a range of environmental threats and the potentially protective role of antioxidants, not only within organisms but also from dietary sources.

- The genetic study of certain human diseases now provides a fuller appreciation of the range of cellular proteins with special functions in the protection, maintenance and repair of gene sequences, as well as the suppression of cancer.

- Our understanding, at a molecular level, of the range of cellular devices in plants, microbes and animals that control the rate and extent to which genes are expressed in response to external and intracellular stimuli is advancing at a spectacular rate.

The topics to be covered in this brief, but broadly based, text are addressed from a fairly general biochemical and molecular biological perspective. Some knowledge of elementary biochemistry and molecular biology is assumed. However, key issues relating to the biochemistry and molecular biology of, for example, DNA replication and gene expression are briefly reviewed. Primarily this is for readers whose knowledge may be fragmentary. A short list of additional reading sources as well as a summary of the key points covered is included with each chapter.

This book is intended for the later stages of undergraduate degree programmes in Biochemistry, Molecular Biology, Biotechnology, Microbiology, Toxicology and Environmental Biosciences. The growing importance attached by governments and health authorities to environmental issues makes their consideration from a molecular biological standpoint an important topic for such students. The book provides an introductory framework around which to integrate critical topics such as DNA damage and repair; gene structure and expression; metabolism of foreign compounds and gene activation; genetic engineering and biotechnology; free radicals and antioxidants; as well as aspects of human cancer and nutrition.

The integrative approach will also be suitable for undergraduates in the later stages of other types of health- or life-science degree course, where they are introduced to elementary biochemistry and molecular biology in their initial years. Moreover, the book will serve as a useful introductory overview for medical students, as well as for postgraduate students and researchers in the life-sciences area.

Roy H. Burdon
Glasgow

CHAPTER ①

Introduction: Genes and the Impact of the Environment

KEY POINTS

- An organism's complement of genes (*the genotype*) sets its potential. However, the properties actually realized (*the phenotype*) depend on the interactions of all these genes with the environment.

- Genes do not operate in isolation. They are components of complex regulatory networks within cells.

- In relation to gene activity there are many layers of feedback regulation from both the physiology of the organism itself and the external environment.

- Certain types of stressful environmental conditions can activate stress genes to produce stress proteins that enable organisms to tolerate such stresses.

- Far from being stable and constant, genes themselves are subject to changes, such as rearrangement and multiplication, as well as to chemical changes resulting not only from the intrinsic instability of DNA, but also as a result of damage from environmental factors.

- There are mechanisms of gene transfer between organisms: genes within the environment are not necessarily limited to particular organisms.

Curiosity regarding the environmental influence on heredity has existed for the past 200 years or so. Indeed, some believed that the environment acting over long periods of time could somehow change genes and specifically direct the course of evolution. One such was Lamarck (1744–1829), who believed that those parts of an organism that were used extensively to deal with the environment became larger and stronger, while those not used deteriorated. One example he cited was the giraffe stretching its neck to new lengths in pursuit of leaves to eat. He proposed that such modifications acquired during the lifetime of an organism were inherited by the offspring. The giraffe's long neck, he argued, was the cumulative product of a great many generations of ancestors stretching higher and higher. However, there is no evidence to support the view that acquired characteristics can be inherited. Nevertheless, in the 1920s, Lysenko, who during the reign of Stalin was allowed to exercise absolute political control over Russian genetics, decreed that geneticists there should accept the dogma of acquired characteristics. In particular, he was concerned with the effects of 'vernalization' in plants, which he believed could be inherited. He proposed that winter varieties of cereals could be genetically converted to spring or summer varieties by cold treatment. We now know that none of the genes involved were altered by the low-temperature treatment. Instead the prior cold treatment merely substituted for the natural chilling period of winter, which is a necessary requirement for stem elongation and flowering of these winter varieties. While it is now accepted that the environment cannot direct evolution, it is nevertheless clear that the activity of many genes can be profoundly influenced by environmental factors, including those that can alter the chemistry of DNA.

1.1 Genes, Genotypes and Phenotypes

The DNA of an organism is a very long double-stranded helical molecule made up of a linear series of molecular units called nucleotides, of which there are four types. *Genes* comprise specific segments of this DNA, and the sequence of nucleotides within a gene constitutes a code which specifies the order of amino acids in the proteins that are the building blocks of tissues and cells, as well as the enzymes and hormones that allow metabolism to proceed. Each gene specifies the construction of a different protein (Chapters 7–11). However, a problem arises if the nature, or arrangement, of the nucleotides in a gene is altered. An aberrant amino acid sequence may be specified in the particular protein that arises from such a modified, or *mutated*, gene, and, in turn, this variant protein may have serious consequences for the organism (Chapter 8).

Nucleotide sequences in DNA, however, do not all encode proteins. Many sequences lying outside the coding segments of genes form part of the machinery that can adjust the activity of specific genes (Chapter 7). Not only do nucleotide sequences in the DNA of an organism determine precisely the type and structure of proteins that an organism will make, but they also comprise part of the signalling machinery that controls the level to which these proteins are produced by the organism in response to environmental conditions (Chapters 7 and 9).

A third important feature of DNA molecules is their ability to serve as templates for the manufacture of copies of themselves. When cells divide and when egg and sperm cells are produced, each progeny cell has a complete set of genes which are virtually identical to the set in the original cells. Although not completely error-free, this DNA replication process forms the basis whereby nucleotide sequence variations within genes and related DNA segments are inherited (Chapter 3).

While an organism's complement of genes (_genotype_) sets its potential, the properties actually realized (_phenotype_) depends on the interactions of all of these genes with the _environment_. An organism's phenotype unfolds during development and maturation when genes, and the products derived from them, interact with one another and with environmental factors. A trait cannot be attributed solely to genes or to the environment, since an organism requires an interplay of both for successful survival. Genes provide the initial guidelines for the development of an organism, and a range of possible phenotypes. Within that predetermined range, a specific phenotype is moulded by environmental influences.

When one is dealing with very complex characteristics such as behaviour, the interactions between heredity and environment are far too complicated to be understood at this time. Nevertheless, there are inherited traits that have a behavioural component as part of their phenotype. The following are examples.

- _Lesch-Nyhan syndrome_, due to a mutant version of a gene involved in controlling the metabolism of DNA building blocks, results in sufferers compulsively biting and mutilating their lips and fingers.
- _Porphyria_, which stems from a defect in the production of haem, the iron-containing part of the oxygen-carrying haemoglobin molecule, gives rise to symptoms which include neurological problems.
- _Fragile X syndrome_, caused by a mutated region near the tip of the long arm of the X chromosome, is associated with mental retardation.

- Mental retardation is also a characteristic of a serious single-gene disease of humans known as _phenylketonuria_. The amino acid phenylalanine cannot be broken down normally. It accumulates in body fluids and can be metabolized to several compounds which can cause brain damage.

1.2 _Gene Activity and the Environment_

Even if a gene is present in an individual organism or cell, it may not be active, or 'switched-up'. Moreover, the number of copies of specific proteins produced per unit time by different genes varies to satisfy cellular requirements. Many gene products are needed only under certain conditions, and regulatory mechanisms that function like an on-off switch allow such products to be made only when required. Other more refined mechanisms can make minor adjustments in the intracellular concentration of a particular protein in response to needs imposed by the environment. Also, when a gene is active, its protein product has the potential to interact with other elements within the environment of the cells as well as with external environmental factors such as dietary components and temperature, examples of which are given below.

- _Intracellular environment_: Sex hormones are one example of agents which can have a profound effect within the cellular environment. Not only can different hormones activate different genes, but also the products of genes can find themselves in two distinct chemical environments depending on the sexual origin of the cell in which the gene is active. This leads to sex-influenced inheritance of conditions such as baldness, which requires not only a specific gene, but also the presence of male hormones.

- *Dietary components*: In early classic studies on the nutritional requirements of the bacterium *Escherichia coli*, the genes responsible for the metabolism of the sugar lactose were found to be fully activated only when lactose was added to the growth medium. In contrast, other bacterial nutritional studies demonstrated that an excess of the amino acid tryptophan in the growth medium could repress the activity of bacterial genes responsible for the manufacture of that amino acid. Nutrient starvation of *Bacillus subtilis* can induce gene activation necessary for sporulation. In mammals dietary components can also have profound effects on gene activation. For example, glucose can promote the activation of genes required in the metabolism of carbohydrates and fats.

- *Temperature*: Siamese cats and Himalayan rabbits provide examples of variations in gene expression caused by environmental temperature. Such cats and rabbits have two copies of a mutant gene for coat colour which results in light fur over all the body except at the extremities. The tail, ears, paws and nose area have the normal dark fur. This development of dark fur is, however, temperature-dependent: the lower the temperature, the darker is the fur. It appears that the enzyme that catalyzes the formation of the normal dark fur pigment is partly defective in these animals and works best at the somewhat lower temperatures that occur away from the main body area.

1.3 *Genes and the Tolerance of Environmental Stress*

Genomic responses to certain stressful environmental threats are now well recognized. There is a wide variety of mechanisms by which specific genes can be activated to yield *stress proteins* (Chapters 9 and 10) used by organisms to *tolerate* specific environmental challenges such as excessive temperature shifts; strong sunlight and ultraviolet radiation; high salinity and drought; toxic metals and chemicals; pathogens and oxidative stress (see Figure 1.1 and Chapters 9–11). The detailed molecular biological mechanisms by which such *stress genes* are activated to counteract environmental adversity are now the subject of intensive study. The mechanisms are likely to provide paradigms for future attempts to explain the more complex interactions between heredity and the environment. Knowledge of these stress genes and the mechanisms of their activation can also have useful applications, for example as early 'indicators' of 'stress' in organisms in threatened environmental locations, or possibly in the genetic engineering of more environmentally tolerant, economic crops (Chapter 10).

1.4 *DNA Stability and the Environment*

While 'switching up' certain stress genes to cope with environmental adversity will contribute to an organism's ability to adapt to changing environments, other mechanisms exist. Animal cells in culture can achieve resistance to toxic drugs by multiplying the number of certain genes. For instance, the gene

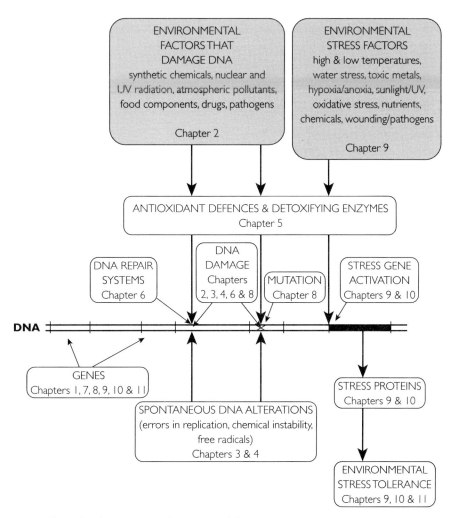

FIGURE 1.1 *Interrelationships between genes and environmental factors*

for dihydrofolate reductase is dramatically multiplied in cultured mammalian cancer cells resistant to the anticancer drug methotrexate. These cancer cells appear to become resistant to the drug by manufacturing a large excess of the protein that normally serves as a target for the drug, thereby diminishing its effectiveness and allowing the cells to thrive even in the drug's presence. Whether this type of *gene amplification* generally occurs in whole organisms is not yet clear. However,

there are indications from studies on herbicide resistance that this may be the case in certain plants.

There are also mechanisms whereby the nucleotide sequences within the DNA of an organism can be rearranged. *Transposable elements* can not only restructure the genome of an organism by their movement, but also cause deletions and amplifications of neighbouring nucleotide sequences in the process. The presence of these mobile genetic elements has been detected in bacteria, plants and certain animals, but in plants and insects it appears that the frequency with which some transposable elements move can be influenced by environmental temperature. Within an environmental context genes can also be 'shared' between organisms. This can occur for instance by *plasmid transfer* among different bacterial species and yeasts, as well as between certain bacteria and plants (Chapters 9 and 10). Other mechanisms of gene transfer include *bacteriophage transduction* and even the direct uptake of DNA by certain bacteria in *transformation* processes (Chapter 11).

A further threat to the stability and constancy of an organism's genes lies in the intrinsic *chemical lability of DNA* itself, which can result in *spontaneous DNA alterations* (see Figure 1.1 and Chapters 3 and 4). Additionally, the generation of *free radicals* within cells as by-products of normal metabolism can contribute to the apparently spontaneous alteration of DNA structures (Chapter 4).

Superimposed on this background of spontaneous change is the ability of many *environmental factors* to cause *specific localized damage* to the chemical structure of DNA (see Figure 1.1 and Chapter 2). Among environmental factors that can damage the DNA of genes and lead to unwelcome mutations are some widely used synthetic industrial chemicals; pollutants in our air, water and food; and sunlight and ionizing radiation. Fortunately, all organisms are equipped with a variety of protective mechanisms such as antioxidant defences and detoxifying enzymes (see Figure 1.1 and Chapter 5), as well as enzymic systems to repair many different types of DNA damage (see Figure 1.1 and Chapter 6). Nevertheless, depending on the nature and severity of the hazard, such systems of protection and repair can sometimes be overwhelmed, or even become damaged themselves. If the DNA of the gene is replicated without the necessary repair having first taken place, an aberrant, or mutated, version of the gene will be inherited by subsequent cell generations (see Figure 1.1 and Chapter 8). In the case of germ cells, which give rise to eggs and sperm, this may have possible long-term consequences not only in terms of inherited diseases, but also in relation to mechanisms of adaptation, survival and evolution. Mutations within genes of somatic cells, which comprise the remainder of the body, are also serious. In the short term they can lead to degenerative conditions such as cancers (Chapter 8), which present immediate and worrying problems for human health and welfare.

It is important to understand the exact chemical nature of the particular DNA damage caused by different environmental factors. We also need to appreciate the scope of cellular protection and DNA repair mechanisms available to different organisms to estimate their potential to tolerate specific environmental adversities. This will help us to devise future strategies for the least toxic environmental practices and for the best protection of living organisms.

1.5 *Scope of the Book*

Figure 1.1 provides a diagrammatic summary of the interrelationships between genes and environmental factors encompassed in the chapters that follow.

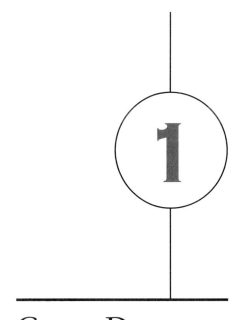

GENE DAMAGE:

Its LIMITATION, REPAIR

and CONSEQUENCES

Environmental Factors that Damage DNA

- Much current endeavour concerns the analysis of the precise nature of the chemical damage to DNA brought about directly or indirectly by various environmental factors

- The environmental threats highlighted include toxic synthetic chemicals; air pollutants including those from natural and industrial sources and automobile exhausts; tobacco smoke and asbestos; the growing problem of ultraviolet radiation as a result of the shrinkage of the ozone layer; ionizing radiation from natural sources and in relation to energy generation; and drugs and components of foodstuffs.

- A comparison is made with other environmental agents such as microbial and viral pathogens which have the potential to alter radically parts of the nucleotide sequence of their host's DNA by complex genetic mechanisms.

Fundamental to an appreciation of the impact of the environment on genes is an understanding of the wide range of different types of damage that can be inflicted on the DNA of organisms following their exposure to environmental hazards. Over many years considerable effort has gone into analyzing precisely the chemical nature of such damage. The aim of this chapter is to provide an introduction to the different types of structural damage to DNA that can be brought by environmental factors which have particular relevance to our present way of life. As indicated in the previous chapter (Figure 1.1), these include toxic synthetic chemicals of major concern in industrial communities; air pollutants from industrial sources and automobile exhausts; the components of tobacco smoke and asbestos; the growing hazards of ultraviolet radiation resulting from the shrinkage of the ozone layer, ionizing radiation from natural sources and energy generation systems; and 'therapeutic' drugs and components of our food. While these environmental hazards can bring about specific structural damage to DNA molecules, a comparison is made with environmental microbial and viral pathogens which have the ability to alter radically segments of their host's DNA, but by distinct complex genetic mechanisms.

2.1 *Synthetic Chemicals*

Over the past 200 years or so human activity has created a new chemical environment. In the middle of the eighteenth century steam power became the dominant energy source of industrial civilization. Carbon-based fuels such as coal, peat, oil and natural gas are burned in large quantities, releasing complex chemical mixtures into the air, water and soils. These fuels are also used by the petrochemical industry to produce the raw materials for synthetic consumer products such as plastics, paints, dyes, pharmaceuticals, cosmetics, food and food additives. Not only does each process release waste chemicals into the environment, but, once used, the products themselves are excreted, dumped or burned, also contributing to the overall load of synthetic chemicals in the environment. Such chemicals also arise during aluminium, iron, steel and coke production as well as in rubber manufacture. Another source has been the industrialization of agriculture, necessary to support the growth of urbanized industrial societies. This latter activity is highly dependent on the use of fertilizers and pesticides at high levels for intensive cultivation, as well as the extensive use of pharmaceutical products in livestock husbandry. Such practices unfortunately also contribute to high levels of synthetic chemicals in water and soil, together with residues in the food chain.

The first significant studies on the specific interaction of synthetic chemicals with DNA arose from the notion that man-made chemicals might prove to be useful lethal and injurious agents in warfare. In recent times, however, we have become seriously concerned with the role that manufactured chemicals can play in the processes which lead to the development of major human diseases, particularly cancer. Many industrial processes that are known to give rise to occupational cancers involve exposure of workers to specific synthetic chemicals such as polycyclic hydrocarbons in aluminium, iron, steel and coke production, and aromatic amines in the rubber and dyestuffs industries. As a consequence, there is now an intense interest in the molecular mechanisms by which different synthetic chemicals can interact with, and damage, DNA. In turn, this has led to a vast literature covering a wide variety of damage to DNA brought about by many types of synthetic chemical. Inevitably, the molecular mechanisms of interaction with DNA have been better studied for some chemicals than others. A few examples are given below.

2.1.1 SOME CHEMICALS CAN ALKYLATE DNA

Monofunctional alkylating agents are chemicals with single electrophilic alkylating groups (e.g. methyl, ethyl) that can interact covalently with single nucleophilic centres in DNA. Although potential reaction centres exist in all four DNA bases, not all of these are equally reactive. The most reactive centres in the base adenine are the nitrogens at positions 1, 3, 6 and 7 (see Figure 2.1 for numbering of atoms). Correspondingly in the case of guanine, the most reactive centres are the nitrogens at positions 2, 3 and 7, as well as the oxygen at position 6 (see Figure 2.1 for numbering). In both bases, however, the nitrogens at positions 3 and 7 are by far the most reactive; Figure 2.1 illustrates the structure of adenine methylated in position N-3 and guanine in position N-7. In the case of the pyrimidine

bases, the most reactive centres in cytosine are N-3, N-4 and O-2, and in thymine N-3, O-2 and O-4. Overall, however, the relative reactivity of all these centres can be extensively influenced by the nature of adjacent bases in the DNA molecule through charge and steric effects. Examples of monofunctional alkylating agents include dimethylnitrosamine, diethylnitrosamine, methyl methane sulphonate, ethyl methane sulphonate, N-ethyl-N-nitrosourea, N-methyl-N-nitrosourea, 1,2-dimethylhydrazine and vinyl chloride (see Figure 2.2). As an example of the kind of reaction which can occur, the interaction of ethyl methane sulphonate with the O-6 and O-4 atoms of guanine and thymine respectively is shown in Figure 2.3.

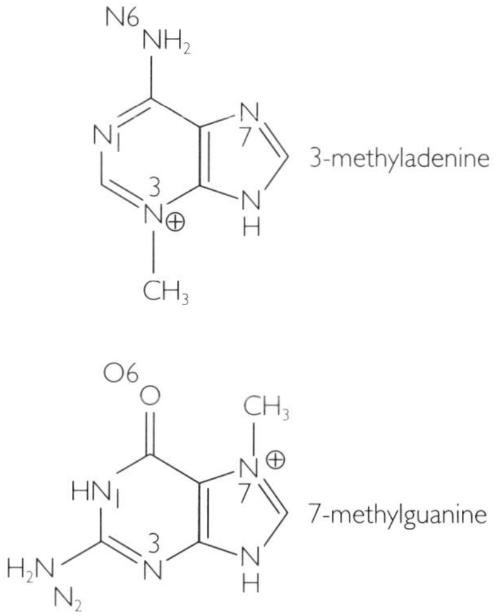

Bifunctional alkylating agents have two reactive groups and in principle can therefore interact covalently with two sites on a DNA molecule. If the two sites are on opposite DNA nucleotide strands, *interstrand* cross-links arise, whereas if the two sites are on the same DNA strand the result is an *intrastrand* cross-link. Interstrand links are particularly problematic as they prevent DNA strand separation and cause toxicity due to the resultant blockage to DNA replication and subsequent cell division. For this reason some bifunctional alkylating agents have been widely utilized as therapeutic drugs in various cancer treatments (see section 2.6). Examples of bifunctional alkylating agents that interact with DNA bases include sulphur and nitrogen mustards, *cis*-platinum(II)diaminodichloride and cyclophosphamide (see Figure 2.2).

2.1.2 SOME CHEMICALS ARE METABOLIZED TO DAMAGING ELECTROPHILIC REACTANTS

Although some *polycyclic aromatic hydrocarbons* such as 7-bromomethylbenzo[*a*]anthracene

FIGURE 2.1 *Alkylation products of DNA bases adenine and guanine*

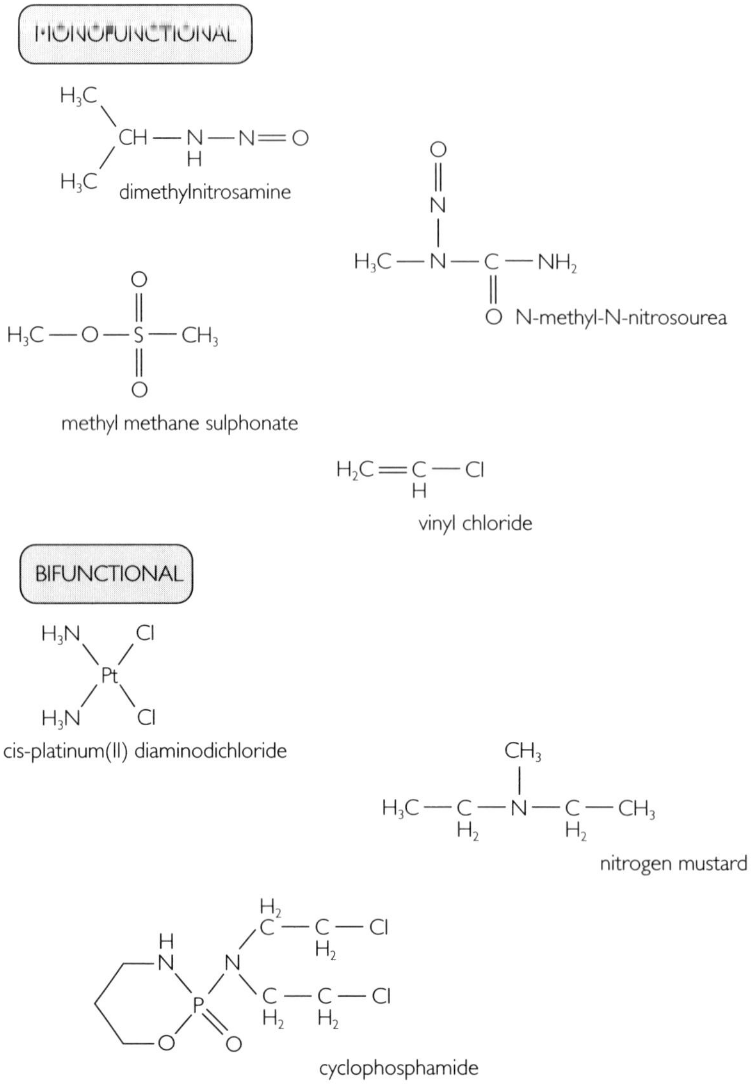

FIGURE 2.2 *Some chemicals capable of alkylating DNA*

can interact directly with the N-6 and N-2 of DNA adenine and guanine bases respectively, many other polycyclic aromatic hydrocarbons require prior metabolism within cells. In fact, it is now widely recognized that many relatively unreactive non-polar environmental chemicals are involved in metabolic transformations catalyzed by enzymes within human and other animal cells, which give rise to more reactive forms that can interact with nucleophilic centres in cellular DNA mole-

cules. Paradoxically, the biological function of these enzyme systems is to protect cells against the cytotoxic effects by converting potentially toxic non-polar chemicals to water-soluble excretable forms. Although this is normally the case, some of these chemicals unfortunately can become activated to electrophilic forms which are notably highly reactive with nucleophilic centres in DNA (Figure 2.4). The process of activation has been associated with a series of membrane-bound

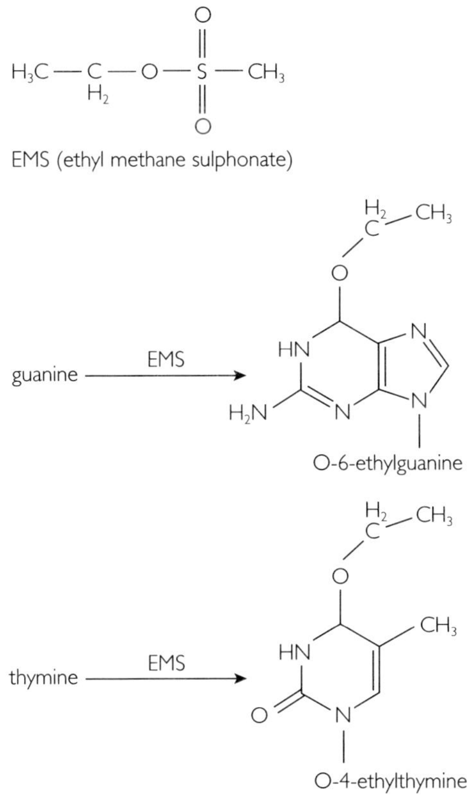

EMS (ethyl methane sulphonate)

guanine —— EMS ——→ O-6-ethylguanine

thymine —— EMS ——→ O-4-ethylthymine

FIGURE 2.3 *Reactions of ethyl methane sulphonate with guanine and thymine*

proteins with monooxygenase activities in combination with one or more membrane-bound flavoprotein reductases, usually referred to as the *cytochrome P-450 system*. For activity,

this multicomponent system, localized in the endoplasmic reticulum, requires NADPH and molecular oxygen. Normally it catalyzes the conversion of hydrophobic non-polar chemicals to more polar oxygenated intermediates and other products. These subsequently serve as substrates for other enzymes such as *glutathione transferase* that catalyze the formation of harmless water-soluble ester conjugates which are excreted (Figure 2.4).

A particularly well studied example of an environmental chemical, hazardous because it can be activated by cytochrome P-450 to a form highly reactive towards DNA, is the polycyclic aromatic hydrocarbon *benzo{a}pyrene*. Originally detected in coal tar, but now also found in cigarette smoke and car exhaust fumes, benzo[a]pyrene is a non-polar planar compound. Components of the cytochrome P-450 system can metabolize it to phenols and dihydrodiols which together with their corresponding ester conjugates can be excreted safely. Unfortunately, in addition to such products, some electrophilic epoxides are also produced. One of these, the diol epoxide 7b,8a-diol-9a,10a-epoxy-7,8,9,10-tetrahydrobenzo[a]pyrene, binds specifically through its C-10 position to the 2-amino position of certain guanine bases in DNA (Figure 2.5), aligning itself in the minor groove of the DNA molecule in the process.

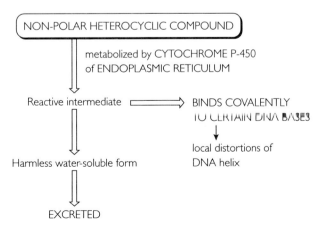

FIGURE 2.4 *The metabolism of non-polar polycyclic aromatic compounds by cytochrome P450 in a mammalian cell*

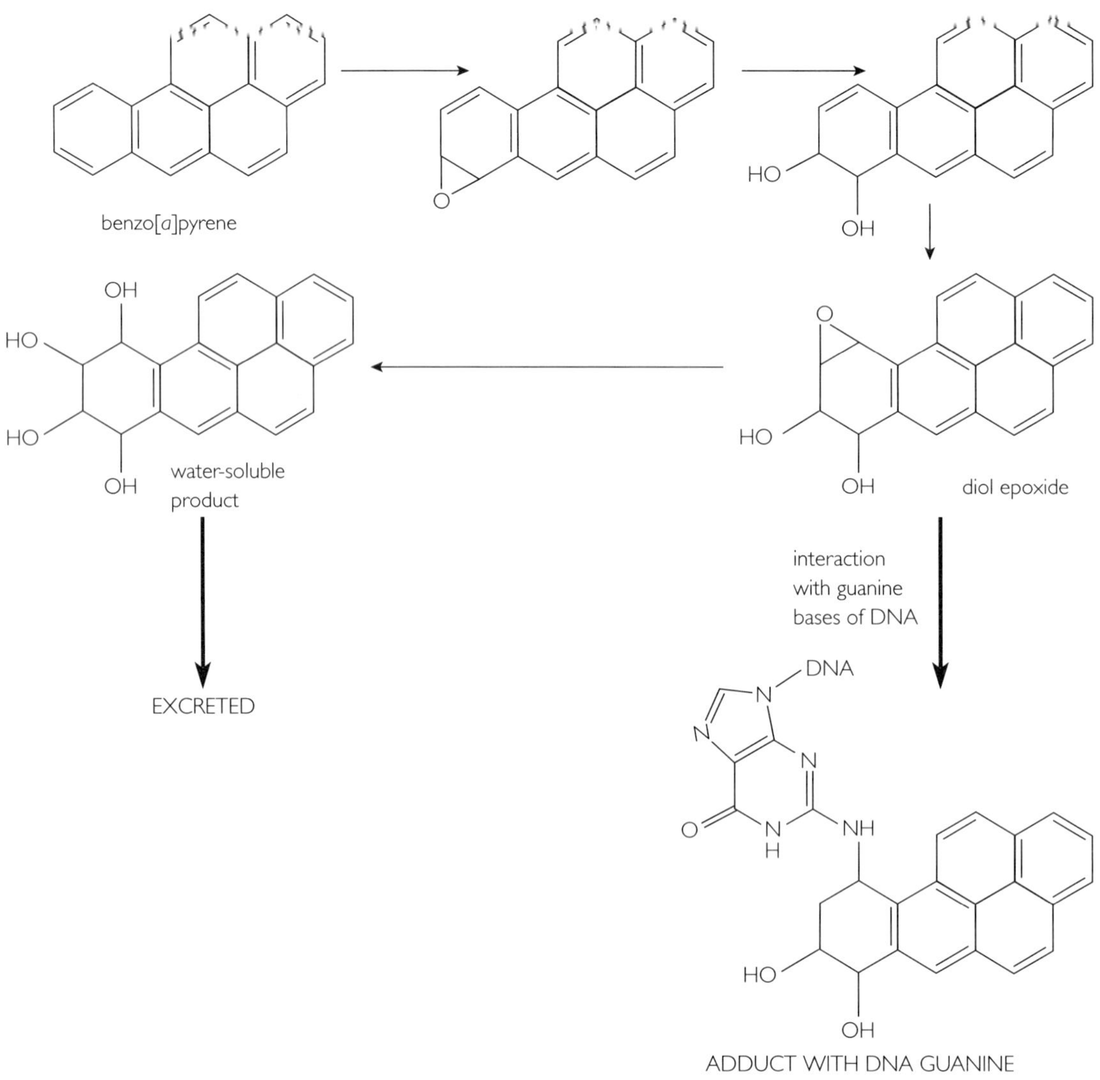

FIGURE 2.5 *The cellular metabolism of benzo{a}pyrene and adduct formation with DNA guanine*

Another notable environmental chemical, hazardous because of its metabolic activation, is the aromatic amine *N-2-acetyl-2-aminofluorene (AFF)*, originally developed as an insecticide. It is acted upon by cytochrome P-450 systems to yield an N-hydroxy derivative. This can subsequently be converted by cellular enzymes such as *sulphotransferases*, or *acetyltransferases*, to highly reactive electrophilic sulphate, or acetate, esters which react readily with DNA guanine residues at the C-8 or N-2 positions (Figure 2.6).

Compounds with structures similar to AFF have recently been found to arise in meat after cooking.

Prior activation by N-hydroxylation and esterification within cells also precedes the covalent interaction of the azo-dye, *4-methylaminazobenzene* through its N atom with the C-4 of DNA guanine residues, whereas the highly carcinogenic polycyclic hydrocarbon *7,12-dimethylbenzo{a}anthracene* is first metabolized to the 5,6-oxide which then interacts with the N-2 of guanine.

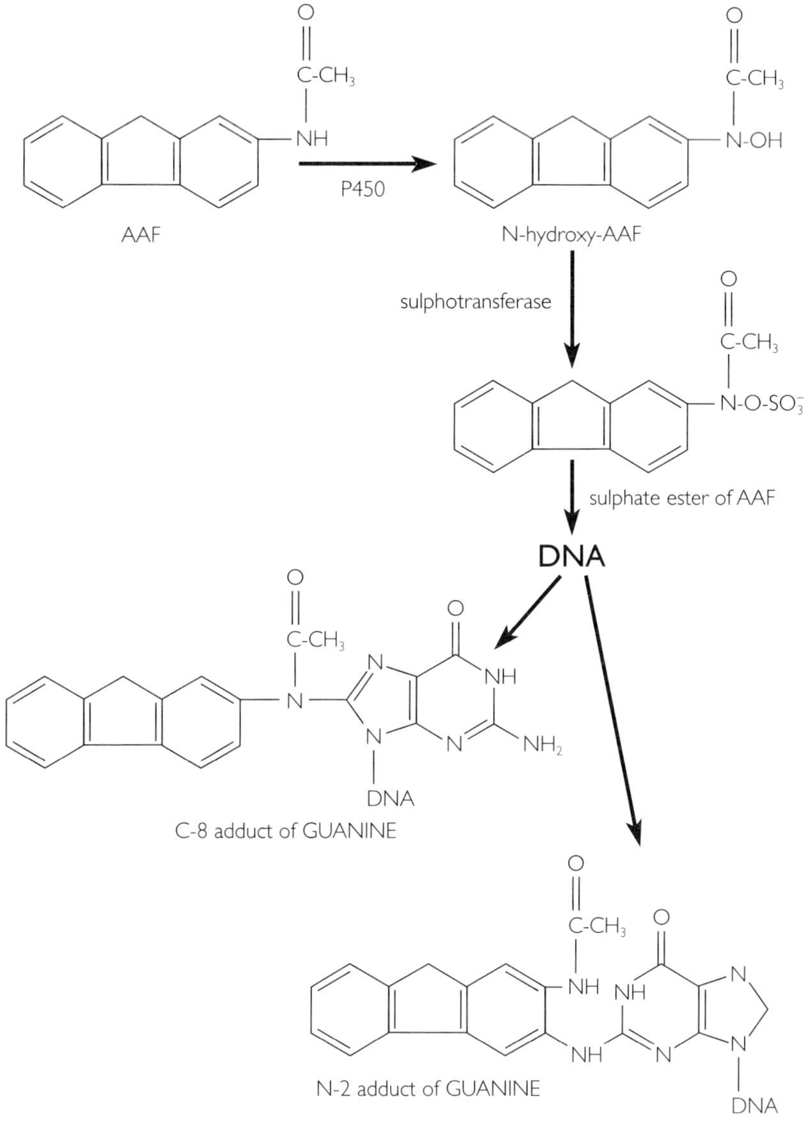

FIGURE 2.6 *Activation of 2-acetylaminofluorene (AFF) and formation of DNA adducts*

2.2 *Nuclear Radiation*

The release of nuclear radiation into the environment is another product of modern civilization. Large-scale releases of radioactivity resulting from the testing of nuclear bombs are no longer considered to be a significant problem. However, in the 1970s there was a growing dependence in a number of countries on nuclear fission as a source of power for the generation of electricity. Public opinion has subsequently emerged which has seriously questioned the safety of the nuclear industry, limiting its growth.

Radioactivity is the emission of energy or subatomic particles (or both) from the spontaneous decay of an unstable atom to a lighter element. There are many naturally occurring radioactive elements and also many synthetic ones that do not occur in nature. At the present time, the average total radiation dose that that most people are exposed to is around 2.2 millisieverts per year. Of this about 10% comes from cosmic radiation from space and around 14% from ground uranium. Medical X-rays account for approximately 12%; nuclear plant emissions and atmospheric bomb testing now only amount to about 0.5%. A present concern, however, is radon, which in certain situations can represent as much as half the annual exposure and may account for as many as 1 in 20 lung cancer deaths in the UK. This colourless, odourless and tasteless radioactive gas results from the decay of naturally occurring uranium in rocks and soil, and can seep into the basement areas of buildings.

The basis of nuclear power is a chain reaction of nuclear fissions of unstable uranium, thorium, or plutonium isotopes in an enclosed nuclear core, or pile, of nuclear fuel. Although nuclear reactors are built with multiple layers of safeguards and the release of radioactivity into the environment from such reactors is usually extremely small, large quantities of radioactive waste can be produced. This has raised serious problems with regard to its handling and long-term storage. Some of the radioactive isotopes produced will remain hazardous for hundreds of thousands of years.

The α, β and γ radiations released in the nuclear decay processes carry a great deal of energy and are dangerous because they can remove electrons away from other atoms with the resultant formation of ions; hence the term *ionizing radiation.* Another form of high-energy electromagnetic radiation capable of ionization is X-rays, which are given off when some of the electrons in orbit around a nucleus jump down from a higher to a lower energy state. The penetrating power of these different forms of radiation varies considerably. The electromagnetic waves of X-rays (and the related γ-rays) are extremely energetic and it takes several centimetres of lead shielding to block their passage. The subatomic β-particles, however, can barely penetrate skin, and α-particles can easily be stopped using a sheet of paper.

Ionizing radiation is particularly hazardous to organisms because of its lethal effects at high levels of exposure. In the 1930s, experiments with fruit flies showed that X-rays were powerful agents of gene mutation, and many early workers with X-rays died of cancer. Sub-lethal exposure in humans to ionizing radiation can induce 'radiation sickness' (skin burns, hair loss, nausea, loss of immune function), but even lesser exposures can cause damage to bone marrow, spleen, lymph nodes and blood cells as well as long-term problems such as infertility, neurological damage and cancer.

Since approximately three-quarters of the mass of most cells is water, around 75% of the ionizations produced within cells by ionizing radiation occur in water molecules. Subsequently the unstable ionized molecules undergo reactions, either in collisions with other molecules or spontaneously, to yield stable, but highly reactive, entities called free radicals. These are species capable of independent existence that contain one or more unpaired electrons. Ionizing radiation can damage the molecular structures of cells in two ways: firstly by direct action where the cellular macromolecules are themselves ionized, or secondly by indirect action, where water or solute molecules are ionized and then generate free radicals, which subsequently damage cellular macromolecules. The precise proportions of direct and indirect damage within cells are difficult to assess, as some cellular macromolecules may be physically isolated from water, or certain soluble compounds within cells may react preferentially with the free radicals, thus providing a protective screen for particular cellular macromolecules.

The primary products of radiation-induced breakdown of water are hydrogen and hydroxyl free radicals (the dot represents an unpaired electron)

$$H_2O \longrightarrow H^. + HO^. \text{ (hydroxyl radical)}$$

These radicals can be formed directly by dissociation of excited water molecules, or by their ionization

$$H_2O \xrightarrow{\text{ionization}} H_2O^+ + e^-$$

The electrons (e^-) ejected can combine with water molecules to yield negatively charged ions.

$$H_2O + e^- \longrightarrow H_2O^-$$

The above positive and negative ions are unstable, and both can dissociate to generate a stable ion and free radical

$$H_2O^- \longrightarrow HO^- + H^.$$
$$H_2O^+ \longrightarrow H^+ + HO^.$$

The hydrogen (H^+) and hydroxyl (HO^-) ions produced can combine to form water, thus overall the reaction produces hydrogen and hydroxyl free radicals. Since electron capture by water molecules is relatively slow, the electrons produced by ionization of water as shown above may become hydrated and survive for a considerable time. Such hydrated electrons (e^- aq) can nevertheless react rapidly with hydrogen ions to yield hydrogen free radicals

$$H^+ + e^- \text{ aq} \longrightarrow H^.$$

$H^.$, $HO^.$ and e^- aq are highly reactive and can interact with various cellular solute molecules including DNA, proteins and lipids as they diffuse through cells. In this process these primary free radicals generate secondary radicals and other related molecules. In turn such secondary products also have the capacity for further attack on important cellular macromolecules.

Another important consideration is that both $H^.$ and e^- aq can react with *oxygen* to form hydroperoxyl radicals ($HO_2^.$) and superoxide radicals ($O_2^{.-}$)

$$O_2 + H^. \longrightarrow HO_2^. \text{ (hydroperoxyl radical)}$$
$$O_2 + e^- \text{ aq} \longrightarrow O_2^{.-} \text{ (superoxide anion radical)}$$
$$O_2^{.-} + H^+ \longrightarrow HO_2^.$$

Depending on their local concentration, radicals can also react with one another: for example, with the formation of hydrogen peroxide

$$HO^. + HO^. \longrightarrow H_2O_2 \text{ (hydrogen peroxide)}$$
$$H^. + HO^. \longrightarrow H_2O$$
$$HO_2^. + HO_2^. \longrightarrow H_2O_2 + O_2$$

The likelihood that these free radicals react with one another, rather than with solute molecules, is a function of the spacing of the ionizations in the water. If the ionizations, and thus the generation of free radicals, are far apart, they may have little opportunity of interaction with one another before meeting solute molecules. In contrast, with densely ionizing radiations such as α-rays, localized concentrations of free radicals can rise considerably. Consequently, relatively high concentrations of hydrogen peroxide are created around the paths of α-particles.

2.2.1 DAMAGE TO DNA BASES CAUSED BY NUCLEAR RADIATION

The attack of free radicals on other molecules can lead to the abstraction, or addition, of hydrogen and the formation of a new radical.

Abstraction

$$RH + H^{\cdot} \longrightarrow R^{\cdot} + H_2$$

$$RH + HO^{\cdot} \longrightarrow R^{\cdot} + H_2O$$

Addition

$$RH + H^{\cdot} \longrightarrow RHH^{\cdot}$$

$$RH + HO^{\cdot} \longrightarrow RHOH^{\cdot}$$

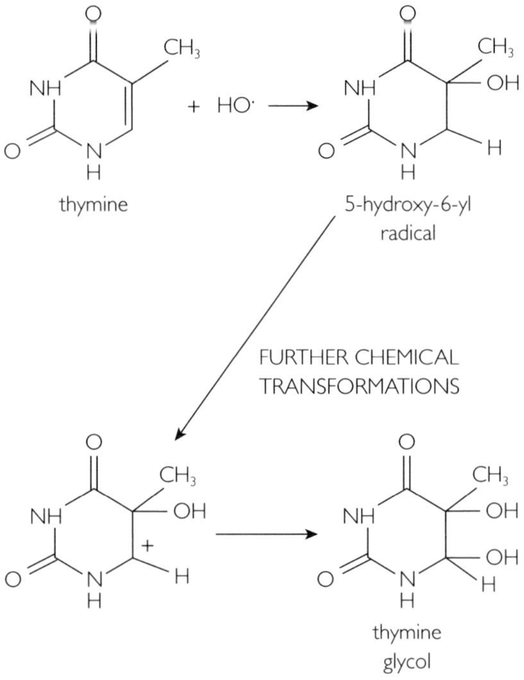

Hydroxyl radicals (HO·), e⁻ aq and hydrogen atoms (H·) can react with the heterocyclic bases of DNA by 'addition'. In the particular case of hydroxyl radicals, these can add to the C5–C6 double bond of pyrimidine bases at diffusion-controlled rates to yield 5-hydroxy-6-yl and 6-hydroxy-5-yl radicals. However, the C5 position of pyrimidines with the highest electron density is the preferred site of attack (Figure 2. 7). 'Abstraction' by hydroxyl radicals of a hydrogen from the methyl group of thymine is also possible.

An important feature is that the radicals formed from DNA pyrimidine bases can undergo further chemical transformations by various oxidation (or reduction) mechanisms to yield a variety of characteristic end-products. By such means the 5-hydroxy-6-yl radical of thymine gives rise to the formation of thymine glycol (see Figure 2.7) and the 5-hydrox-6-yl radical of cytosine yields cytosine glycol.

Hydroxyl radicals can also add to C4, C5 and C8 positions of purine bases. Figure 2.8 illustrates the structures of hydroxyl adducts of guanine. Also shown in Figure 2.8 is the result of further chemical transformations

FIGURE 2.7 *Attack of hydroxyl radicals on DNA thymine to yield thymine glycol*

which yield a characteristic end-product, *8-hydroxyguanine (or its keto form, 8-oxoguanine)*. Similar mechanisms yield 8-hydroxyadenine. Formamidopyrimidines (Fapys) are formed by the one-electron reduction of the imidazole ring-opened forms of purine C-8 hydroxyl adduct radicals. Some characteristic end-products of radical attack on DNA bases are given in Table 2.1, and illustrated in Figure 2.9.

2.2.2 DAMAGE TO DEOXYRIBOSE COMPONENTS OF DNA BY NUCLEAR RADIATION

Hydroxyl radicals sometimes also react to a lesser extent with the deoxyribose components of DNA by extraction of hydrogen atoms from any of the carbon atoms. The extent of this attack on deoxyribose residues

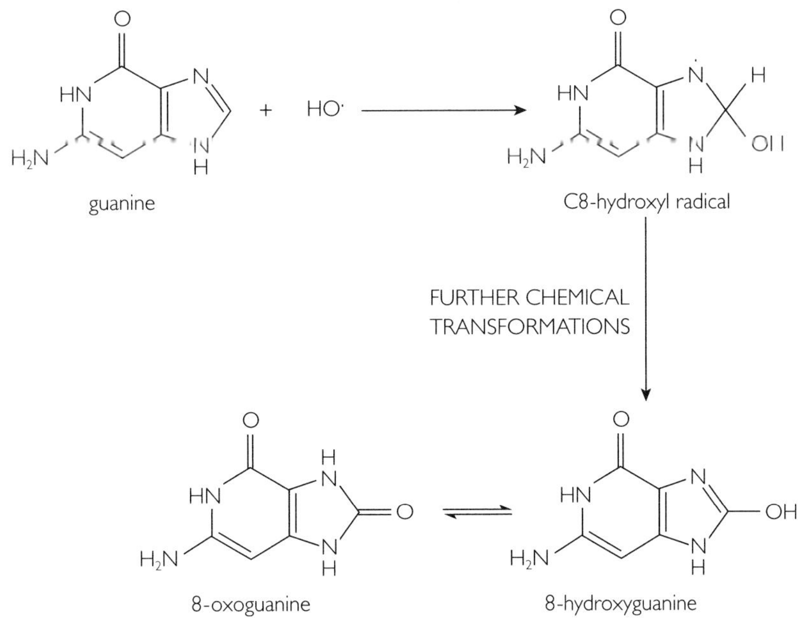

FIGURE 2.8 *Attack of hydroxyl radicals on DNA guanine to yield 8-hydroxyguanine*

TABLE 2.1 *Some products of free radical attack on DNA*

Bases	Deoxyribose
thymine glycol	2,5-dideoxypentos-4-ulose
cytosine glycol	2,3-dideoxypentos-4-ulose
8-hydroxyguanine	2-deoxypentos-4-ulose
8-hydroxyadenine	2-deoxypentonic acid
2-hydroxyadenine	erythrose
5-hydroxycytosine	2-deoxytetrodialdose
5,6-dihydrothymine	glycolic acid
5,6-dihydroxycytosine	
5-hydroxy-6-hydro-thymine	
6-hydroxy-5-hydro-thymine	
5-hydroxy-6-hydro-cytosine	
6-hydroxy-5-hydro-cytosine	
4,6-diamino-1-hydroxy-5-formamidopyrimidine	
4,6-diamino-5-formamidopyrimidine	
5-(hydroxymethyl)-uracil	
5-hydroxyhydantoin	
5-hydroxy-5-methyl-hydantoin	
8,5'-cyclo-2'-deoxyadenosine	
8,5'-cyclo-2'-deoxyguanosine	

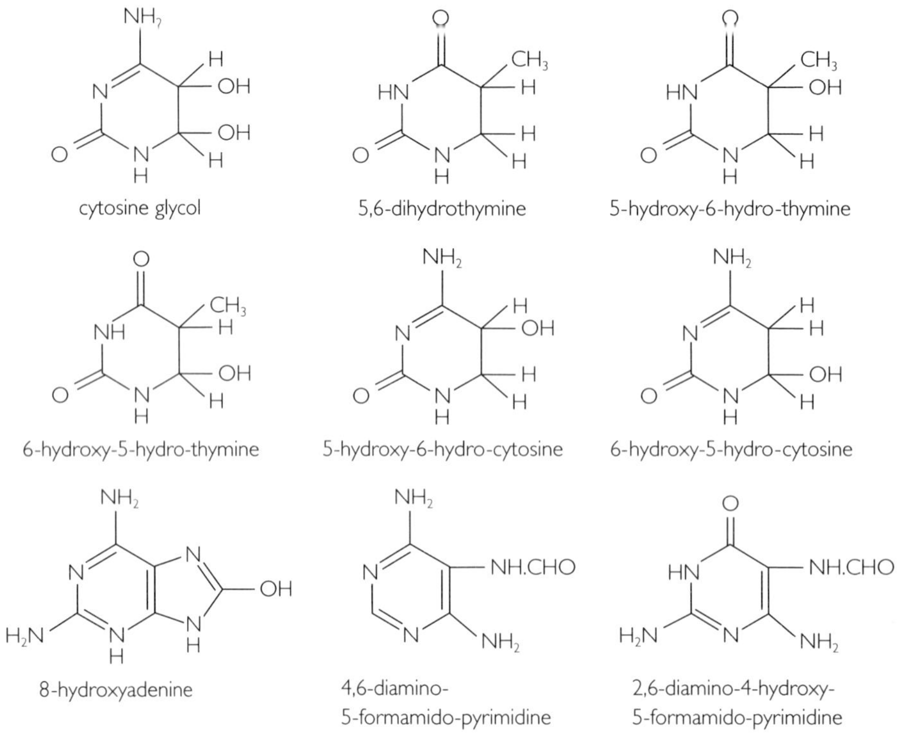

FIGURE 2.9 *Some characteristic end-products of radical attack on DNA*

depends on the single- or double-strandedness of the DNA itself. However, like the DNA base radicals, the deoxyribose radicals formed can subsequently undergo various chemical transformation reactions. For instance, the oxidation of the C4′-centred deoxyribose radical leads to 2,5-dideoxypentos-4-ulose, 2,3-dideoxypentos-4-ulose and 2-deoxypentos-4-ulose. In the presence of oxygen, the cleavage of C–C bonds of sugar peroxyl radicals can also take place. As a result, an erythrose residue within DNA and a glycolic acid residue attached to a broken DNA chain are formed. In general, the altered sugar residues that result from radical attack (see Table 2.1) can either remain bound to DNA with both phosphate linkages still intact, or be released from the DNA chain, with concomitant strand breakage. However, in some cases, as just mentioned, the altered deoxyribose

moiety can remain attached to a fragmented DNA strand by one phosphate bond (see Figure 2.10).

The identification, in hydrolysates of DNA samples from irradiated cells, of the types of modified base and deoxyribose listed in Table 2.1 has in recent years mainly been carried out using the technique of gas chromatography–mass spectrometry (GC-MS). Such modified bases and sugars now serve as specific 'footprints' of previous free radical attack on the DNA under investigation. Rapid methodologies such as HPLC, with highly sensitive electrochemical detection, also exist for quantitation of nucleosides such as 8-hydroxy-2′-deoxyguanosine and thymidine glycol, in enzymatic digests of cellular DNA.

As already mentioned, radical-induced alteration to deoxyribose moieties can often lead to DNA strand breakage. Because DNA has

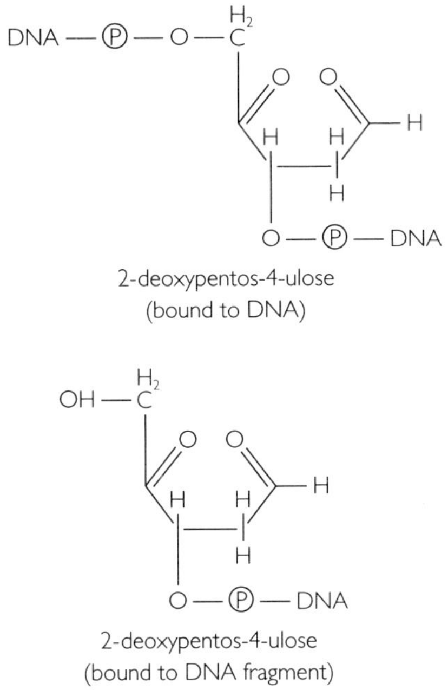

DNA — (P) — O — C (H₂)

2-deoxypentos-4-ulose
(bound to DNA)

OH — C (H₂)

2-deoxypentos-4-ulose
(bound to DNA fragment)

FIGURE 2.10 *Products of radical attack on deoxyribose of DNA*

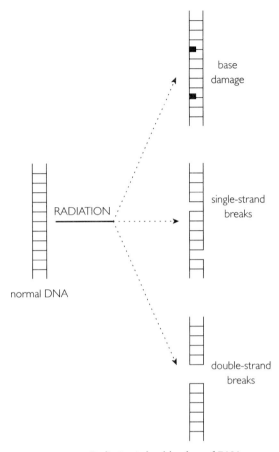

RADIATION

normal DNA

base damage

single-strand breaks

double-strand breaks

FIGURE 2.11 *Radiation-induced breakage of DNA strands*

two strands, breakage of only one of them does not disrupt the molecule as a whole and if randomly distributed, many breaks can accumulate before the DNA is degraded into two or more pieces (Figure 2.11). Thus the free radicals generated from sparsely ionizing radiations probably attack randomly and cleave only one strand at a time, thus generating *single-strand breaks (SSBs)*. In contrast, densely ionizing radiations are more likely to sever each chain at the same or nearly the same point, and thus cause a *double-strand break (DSB)*. A very wide variety of general laboratory techniques are now available for the study and quantitation of strand breakage in isolated DNA samples. These include agarose gel electrophoresis, pulsed-field electrophoresis, alkaline and neutral elution techniques from columns of hydoxyapatite (for single- and double-strand breaks respectively), and ultracentrifugal sedimentation methods. At present there is a large (but controversial) body of opinion that implicates double-strand breaks

as the major lesions leading to cell death after irradiation. In this context it seems that single-strand breaks and base damage are less important.

2.3 *Ultraviolet Radiation*

2.3.1 ULTRAVIOLET RADIATION AND THE OZONE LAYER

Not all types of potentially harmful solar radiation can penetrate the entire atmosphere and reach the earth's surface. Solar X-rays, γ-rays, infrared radiation and most cosmic rays are screened out. Only certain radio frequencies and radiation around the wavelength of visible light, together with some short wavelength infrared and longer wavelength ultraviolet radiation (UV-A and some UV-B) can reach sea level. The short wavelength ultraviolet radiation, UV-C (<280 nm), which is especially biologically hazardous, as it has the correct wavelength to be absorbed directly by DNA, is completely absorbed by the atmosphere.

Long wavelength ultraviolet radiation, UV-A (315–400 nm), is not normally associated with injury to biological systems. Although it does not have a direct interaction with DNA, it can be absorbed by other cellular chromophores, whose excitation can generate potentially harmful reactive oxygen species such as hydroxyl radicals and singlet oxygen. Radiation between 280 and 315 nm (UV-B), however, does pose a significant biological threat as its photons are weakly absorbed by the same molecules that best absorb UV-C and UV-A. Nevertheless, an impediment to UV-B penetration of the atmosphere is the accumulation of a 'layer' of ozone molecules in the stratosphere. This layer is normally around 99% effective in screening off hazardous UV-B radiation.

In the upper atmosphere (thermosphere), 80 km and more from the surface of the earth, oxygen exists predominantly in its monoatomic form (O). This is due to the effects of high-energy photons of sunlight at wavelengths of less than 242 nm on molecular oxygen (O_2) and nitrogen dioxide (NO_2). However, below this level in the stratosphere (15–40 km above sea level), some of these oxygen atoms can combine with molecular oxygen to produce ozone (O_3). Since the 1960s there has been considerable alarm over the fact that ozone levels in the stratosphere over the South Pole have been dropping, giving rise to a seasonal

'ozone hole' This is of critical importance, since for every 5% depletion in stratospheric ozone, it is estimated that there is a 10% increase in UV-B penetration to sea level.

There are two current hypotheses to explain the changes in polar ozone. One, a dynamic model, suggests that the cause could simply be the breakdown of the polar vortex of wind systems in the springtime. The other, which has received more publicity, is a chemical model based on the possibility that in a set of complex reactions, chlorofluorocarbons (CFCs) are broken down by light, or by reaction with atomic oxygen, to yield chlorine atoms. In turn these chlorine ions can each destroy many ozone atoms by forming chlorine oxides (ClO)

$$CCl_3F \xrightarrow{\text{light or O}} CCl_2F + Cl$$
(freon, a CFC)

$$Cl + O_3 \longrightarrow ClO + O_2$$

CFCs, although originally designed for use as non-toxic, nonflammable, chemically inert propellants for aerosol spray cans, are now mainly utilized as refrigerants, solvents for cleaning electronic circuits and in the production of plastic foam products. A particular worry is that each CFC molecule released at ground level takes around 15 years to reach the stratosphere. Even when degraded to chlorine their lifespan there is still considerable, and each chlorine ion can destroy many thousands of stratospheric ozone molecules. For these reasons, there is concern that reductions in CFC use may not be quick enough to protect biological systems from the dangers of increased UV-B radiation.

2.3.2 ULTRAVIOLET DNA PHOTOPRODUCTS

In terms of DNA damage, the most frequent DNA photoproducts made by UV-C and

UV-B radiation join adjacent cytosines or thymines in oxygen-independent reactions. UV-C and UV-B photons of solar light are usually absorbed at the 5–6 double bond, allowing it to open. Where two pyrimidines are adjacent, one of two events usually takes place. If both 5–6 bonds are open, a ring is formed, creating the *cyclobutane dimer* (Figure 2.12A). Alternatively the double bond of the of the 5′ pyrimidine opens and reacts across the exocyclic group of the 3′ pyrimidine. Following spontaneous rearrangement, a single bond is left between the adjacent pyrimidines. This is referred to as the *pyrimidine–pyrimidine(6–4) photoproduct* (Figure 2.12B). Both photoproducts bend the DNA or rotate a base such that it resembles a site lacking a base. In addition to these characteristic dimeric pyrimidine photoproducts, some minor products such as thymine–adenine dimers can also detected.

UV-A radiation of cells can also cause DNA damage, but only *indirectly*. This appears to be due to mechanisms that generate singlet oxygen ($^1O_2$) and require the participation of oxygen together with as yet unidentified endogenous cellular photosensitizers. The singlet oxygen produced by the illumination of these photosensitizers with UV-A (or visible light) can subsequently cause DNA base damage, the formation of 8-oxo-guanine being a notable product, although some strand breaks can also result.

2.4 *Atmospheric Pollutants*

Air pollution is everywhere. Traces of pollutants can be carried world-wide by winds. Although there are natural contaminants, most pollution has arisen since our discovery of fire, but increasing populations and industrialization have now created vast levels of

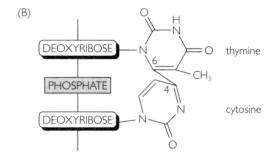

(A)

THYMINE CYCLOBUTANE DIMER

(B)

THYMINE-CYTOSINE 6:4 PHOTOPRODUCT

FIGURE 2.12 *(A) Structure of the cyclobutane thymine dimer formed in DNA by covalent interaction of two adjacent thymines in the same DNA strand; (B) structure of the thymine–cytidine photoproduct between C-6 of a thymine and the C-4 of an adjacent cytosine*

atmospheric pollutants. Primary pollutants are released directly into the air in a harmful form; secondary pollutants are modified to a hazardous form, or are created by physical and chemical reactions in the atmosphere. For instance, some of the most harmful kinds of contaminants are photochemical oxidants created by reactions driven by solar radiation. Overall, the transport, dispersal, concentration and removal of pollutants from the atmosphere is governed by considerations such as topography, adsorption, wind currents, air stability and precipitation.

In terms of their potential to cause specific damage to DNA and thus the possibility of altering the genetic characteristics of cells and organisms, a number of pollutants have been the subject of especial investigation in recent times. These include polycyclic aromatic hydrocarbons, sulphur dioxide, nitrogen oxides, tobacco smoke, asbestos, automobile exhaust fumes and ozone.

2.4.1 POLYCYCLIC AROMATIC HYDROCARBONS

Many polycyclic aromatic hydrocarbons (PAHs) have been identified in the environment. They were originally characterized as pyrolysis products of oils and biological materials, but are also generated in tobacco smoke and by incomplete combustion of fossil fuels such as coal and petrol. Examples include benzanthracene, dibenzanthracene, 12-methylbenzanthracene, 7,12-dimethyl-benzanthracene and benzo[a]pyrene. PAHs form adducts with purine bases, especially guanine, after metabolic activation involving cytochrome P450 enzymes, an example of which is shown in Figure 2.5.

2.4.2 SULPHUR DIOXIDE

Atmospheric sulphur dioxide has irritant properties and causes long-term respiratory problems for humans. It may also possibly contribute to the development of lung cancer. In plants it causes chloroplastidic damage and chlorosis. A major source is the burning of oil and fossil fuel, particularly low-grade coals of high sulphur content. The replacement of coal with cleaner domestic heating fuels has reduced emission in the UK since 1970 by around 60%. There are also natural sources of sulphur dioxide, for example from microbial activity, sulphur springs, volcanoes, and weathering of minerals.

Sulphur dioxide (SO_2) readily dissolves in water and tissue fluids to yield sulphite (SO_3^{2-}) and bisulphite (HSO_3^-) anions.

$$SO_2 + H_2O \rightleftharpoons H_2SO_3 \rightleftharpoons$$
$$H^+ + HSO_3^- \rightleftharpoons H^+ + SO_3^{2-}$$

Both anions have lone pairs of electrons on the sulphur atoms and as a result are readily oxidized in a series of complex overlapping reactions which involve the formation, or consumption, of free radicals. Although most biological tissues have a range of antioxidant defence systems (see Chapter 5) to remove free radicals as they are formed, nearly all organisms have a molybdenum-containing *sulphite oxidase* to remove the major proportion of sulphite.

Despite these protection mechanisms it is possible that sulphite, or free radicals derived from its oxidation (e.g. HSO_3^-, O_2^-, SO_2^-, $HO^•$) can damage DNA bases and bring about chain cleavage. Bisulphite can specifically cause the conversion of cytosine bases to uracil in DNA (Figure 2.13) and is known to be mutagenic

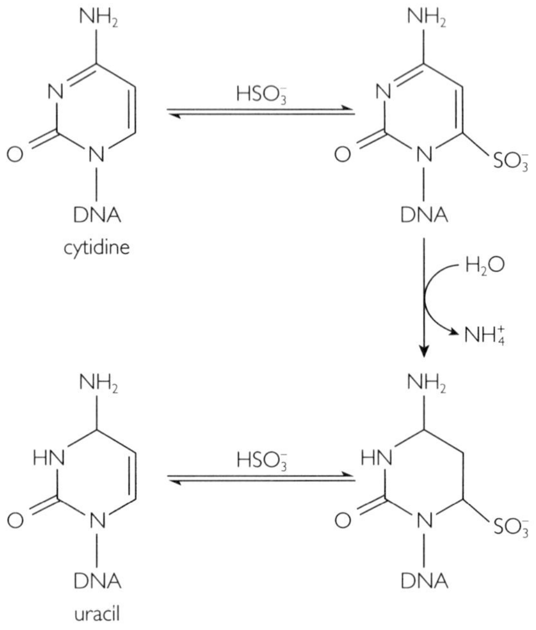

FIGURE 2.13 *Conversion of cytidine to uracil in DNA by bisulphite anions*

to bacteria. On the other hand, the potent microbial sterilizing effect of sulphur dioxide is more likely to be due to its ability to attack disulphide bridges within vital cellular enzymes and structural proteins.

2.4.3 NITROGEN OXIDES

The composition of nitrogen oxides in the atmosphere is complex and they are involved in many different reactions which are influenced by temperature, light and local concentration. Generally nitrogen oxides are formed when nitrogen in fuel or combustion air is heated above 650°C in the presence of oxygen. Principal origins are automobile exhausts, power stations and various industrial processes. Another source is bacteria in soil or water which oxidize nitrogen-containing compounds. The primary product is nitric oxide (NO), but this rapidly oxidizes in the atmosphere to nitrogen dioxide (NO_2). This brownish gas gives photochemical smog its colour, damages vegetation and contributes to human lung tissue damage. Nitrogen oxides also combine with water to give nitric acid, which is a major component of atmospheric acidification.

As yet there is no proof of any mutagenicity, but nitrogen dioxide exposure can nevertheless cause DNA strand breaks in lung cells of experimental animals. This may be due to its ability to react with cellular lipids and generate a variety of reactive free radicals capable of damaging DNA *indirectly*.

2.4.4 AUTOMOBILE EXHAUST

As previously pointed out, ozone performs an essential function in the stratosphere as a shield from life-threatening ultraviolet radiation. In contrast, if present in the lower atmosphere ozone can present considerable problems for plants and animals as it is highly reactive towards organic molecules. Exposure of humans and animals causes lung damage. The problem comes about because one route to the formation of ground level ozone is the photochemical reactions between oxygen, nitrogen oxides and traces of unburnt hydrocarbons in air (such as acetylene, benzene, butanes, ethane, pentanes, propane and toluene). A suitable mixture of these is readily available from automobile exhaust.

In terms of chemical activity, although ozone is not a radical, it can react with a variety of biological molecules including membrane lipids to produce a spectrum of free radicals (Figure 2.14) that can cause pathological damage to plant and animal tissues.

Because components of biological tissues are in close proximity to water, a different type of ozone attack is also possible. Specifically this yields hydrogen peroxide, hydroxyperoxides and reactive aldehydes (e.g. malondialdehyde) in a process termed *ozonolysis* (Figure 2.14). The hydrogen peroxide produced could possibly interact with free iron within cells in a Fenton-type reaction to generate highly reactive free radical species such as hydroxyl radicals (HO·).

$$Fe^{2+} + H_2O_2 \xrightarrow{\text{Fenton reaction}} HO\text{·} + Fe^{3+} + HO^-$$

Besides their ability to initiate the peroxidative damage of cellular membranes, free radicals generated from ozone may contribute to another important aspect of ozone toxicity, i.e. its ability to damage DNA. When rats are exposed to low levels of ozone, single-strand breaks in DNA are observed in cells isolated from their lungs. *In vitro* studies have demonstrated that exposure of animal cells to ozone will induce chromosomal aberrations as

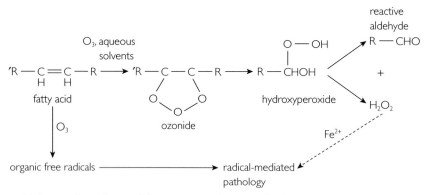

FIGURE 2.14 *Mechanisms for production of damaging species from ozone–lipid interactions*

well as both single- and double-strand DNA breaks.

Automobile exhaust also contains nitric oxide from combustion processes. This can give rise to nitrogen dioxide, which, like ozone, is a component of photochemical smog. These airborne nitrogen oxides can also provide yet a further hazard for cellular DNA. For example, on the surface of microscopic airborne particles in the exhaust fumes they can react with ketone by-products of burning fuel (particularly diesel) to form nitrated polycyclic aromatic hydrocarbons such as 3-nitrobenzanthrone. Together with 1,8-dinitropyrene, also found in diesel exhaust, they are probably the most powerful known mutagens, and appear to be created faster in smoggy air with high levels of nitrogen oxides and ozone. Although the mechanisms whereby they alter DNA nucleotide sequences are not yet clear, they can bring about chromosomal aberrations in experimental mice and could contribute to lung cancer cases in cities. In addition to carbon monoxide, other biologically hazardous components of automobile exhausts include benzene and 1,3-butadiene. Although both of these latter compounds have been linked with increased leukaemia rates, they do not appear to damage DNA directly.

2.4.5 FORMALDEHYDE

Incomplete combustion of fuels and evaporation of solvents release a variety of hydrocarbons into the atmosphere. Although these include acetylene, benzene, butanes, ethane, hexanes, pentanes, propane and toluene, by far the most abundant atmospheric hydrocarbon is methane, which is released from decaying vegetation as well as from industrial and domestic sources. The oxidation of this methane is the major outdoor source of formaldehyde in the atmosphere. In photochemical reactions methane is converted to methoxy radicals ($CH_3O^{\cdot}$), which can subsequently react with oxygen to form formaldehyde (HCHO) and peroxy radicals ($HO_2^{\cdot}$). Although formaldehyde can reach concentrations of around $0.1–1$ nl l^{-1} in the atmosphere, this depends somewhat on climatic conditions and levels of sunlight. Indoors, formaldehyde can be slowly released from formaldehyde resins, or urea-formaldehyde based insulation materials used by the building industry.

Although formaldehyde is an established cause of skin dermatitis, there is concern because it can also cause cancers in experimental animals. It can attack the free amino groups of DNA bases to produce methylol derivatives. In turn these can produce cross-links between two DNA bases, or between DNA bases and amino acids of adjacent proteins

$$BASE^1-NH_2 \xrightarrow{\quad HCHO \quad} BASE^1-NH-CH_2OH$$
$$\xrightarrow{\quad\quad} BASE^1-NH-CH_2-NH-BASE^2$$

2.4.6 TOBACCO SMOKE

Tobacco smoke is perhaps one of the most significant atmospheric pollutants from the human health point of view. At least a third of human cancers are now believed to be tobacco smoke-related. Cigarette smoke is highly complex and contains over 3500 chemicals, of which at least 40 are individually known to cause cancer. These include poly-nuclear aromatic hydrocarbons (PAHs) such as benzo[*a*]pyrene, aromatic amines, aldehydes, and nitrosamines which can form adducts either directly with DNA bases, or indirectly after cytochrome P-450-catalyzed modification, as described for synthetic chemicals in section 2.1. For instance, N′-nitrosonicotine, the most prevalent oesophageal carcinogen in tobacco smoke, forms DNA adducts following metabolic activation by α-hydroxylation. The level of DNA adducts is higher in many tissues of smokers compared with non-smokers. Typically levels of adducts lie in the range of one per 10^{-6}–10^{-8} nucleotides. Experiments with rats after chronic exposure to cigarette smoke suggest that adduct formation in cells of the upper digestive tract can be even higher following ethanol consumption.

Cigarette smoke condensate can itself also generate both superoxide and hydrogen peroxide and can induce DNA strand breaks in cultured animal cells through the generation of these species as the process can be partly inhibited by catalase (which destroys hydrogen peroxide). Extracts from the tar fraction also generate hydrogen peroxide and similarly damage DNA. A possibility is that some of the damage to cellular DNA arises from oxidants, possibly hydroxyl radicals, generated from hydrogen peroxide through Fenton-type reaction involving catalytic iron from cellular sources as described above. Among the many products of DNA damage is the characteristic oxidized base, 8-hydroxyguanine (see Figure 2.8). The tar fraction in each puff of smoke contains around 10^{14} radicals, most of which are semiquinone radicals.

2.4.7 ASBESTOS

Exposure of humans to asbestos can result in asbestosis, diffuse pulmonary fibrosis, or cancers such as bronchiogenic cancer or malignant mesothelioma. Six types of asbestos have been identified: actinolite, amosite, anthphyllite, crocidolite, tremolite, and chrysolite. All contain long chains of silicon and oxygen that give rise to the fibrous nature of the mineral, yet each is different in physical properties depending on other components such as calcium, magnesium or iron. The strength and resilience of asbestos, while important industrially, make it dangerous for human health. Asbestos fibres, particularly of the first five types, can readily penetrate bodily tissues, especially the lungs. *Chrysolite*, which accounts for 95% of the use of asbestos world-wide, may in fact be less hazardous, as it has a slightly different crystal structure, is notably softer, and is more readily broken down in tissues.

It now appears that asbestos fibres can generate highly reactive hydroxyl radicals in the presence of hydrogen peroxide. Pretreatment of the fibres with the iron chelator desferrioxamine inhibits this process and it has been proposed that the hydroxyl radicals are generated through a Fenton-type reaction.

$$\text{asbestos–Fe}^{2+} + H_2O_2 \longrightarrow$$
$$\text{asbestos–Fe}^{3+} + HO^- + HO^{\cdot}$$

Exposure of mammalian cells to asbestos (crocidolite) fibres has been shown to induce DNA strand breaks as well as peroxidation of membranes. Such damage is attenuated if the fibres are first treated with the iron chelator, desferrioxamine. From such observations it is likely that iron-mediated oxidative damage to cell components including DNA could be an important component of asbestos pathogenicity. Fibres with the highest iron content show the greatest propensity to damage DNA, and under certain conditions the oxidized DNA base 8-hydroxyguanine can be detected in the DNA of afflicted cells.

2.5 *Food Components*

Although food is necessary to support life, there are many potentially harmful components of food and several of these are known to cause specific damage to cellular DNA under certain conditions. Food and beverages, a large proportion of which are plant-derived, are by far the major source of chemicals entering the human body. In general plants naturally synthesize an enormous variety of potentially toxic compounds, some of which provide them with vital protection against fungi, insects, and animal predators. Out of 5000 to 10 000 of such 'natural pesticides', 63 have been tested, and of these 35 produced cancers in rodents. Additionally, cooking of food often produces burnt material, some of which can also be carcinogenic to rodents. Roasted coffee contains some 1000 different chemicals and of 28 so far tested, 19 are rodent carcinogens.

On this basis our diet is likely to be a prime source of compounds potentially hazardous to our DNA. A crucial aspect of current debate, however, is whether the high concentrations of apparently harmful dietary components administered to experimental animals to elicit cancers bear any significant relationship to the considerably lower concentrations consumed by humans on a daily basis. With this caveat in mind, several examples are presented below to illustrate some of the specific types of damage to cellular DNA that can be brought about by dietary components.

2.5.1 AFLATOXINS

A particular group of potentially harmful food components discovered in the 1960s is the aflatoxins, which are mycotoxins produced from the fungi *Aspergillus flavis* and *Aspergillus parasiticus.* These particular moulds commonly infest a wide range of foodstuffs, especially nuts and cereals, and the aflatoxins they produce are particularly noted for their ability to cause liver cancers in humans. One particularly potent aflatoxin is aflatoxin B_1 which is oxidized within tissues, such as liver, by the cytochrome P-450 system of the endoplasmic reticulum to yield aflatoxin B_1-8-9-epoxide which preferentially attacks guanine residues in certain nucleotide sequences of liver cell DNA to form a bulky adduct (Figure 2.15).

2.5.2 NITROSAMINES

Nitrite is a meat preservative which acts as an antimicrobial agent, retarding the spore germination of *Clostridium botulinum*. It is also responsible for the reddening effect and the preservation of flavour of cured meats. Despite these desirable effects, nitrite is also the culprit in the formation of mutagenic and carcinogenic N-nitrosamines from amines and amino acids in certain cured meats, such as bacon, under the high temperatures of pan

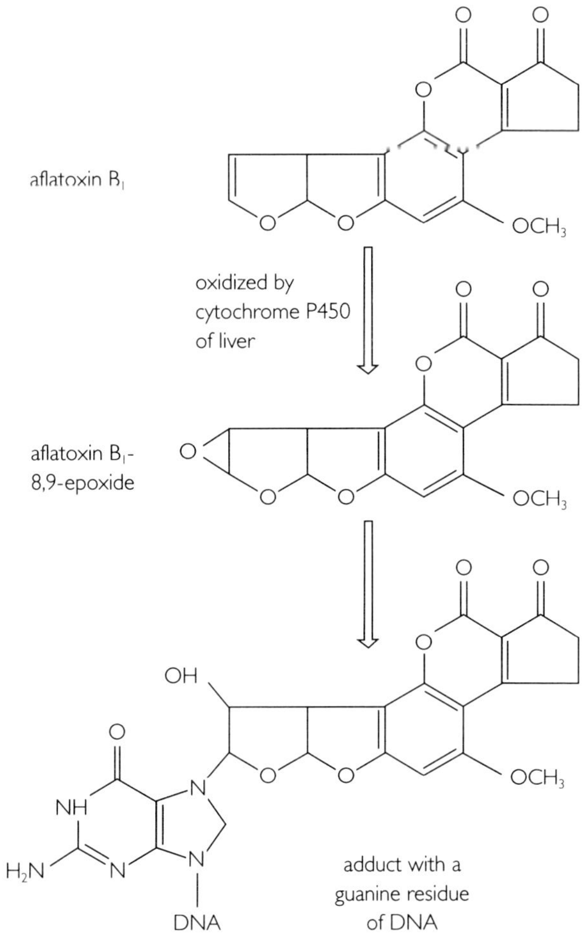

FIGURE 2.15 *Metabolic activation of aflatoxin B₁ and its modification of DNA*

frying. Examples of such products are N-nitrosopyrrolidone, which is structurally related to N-nitrosonornicotine of tobacco smoke, and N-nitrosodimethylamine. Although the molecular mechanism by which such nitrosamines exert their effects at the level of cellular DNA is not yet known, like other mutagenic nitrosamines they may well alkylate DNA bases (see section 2.1).

2.5.3 HETEROCYCLIC AMINES

During the last ten years or so, more than a dozen mutagenic heterocyclic amines have been isolated from fried, broiled, deep-fried or baked meat and fish. Such amines can arise from heating mixtures of creatine, free amino acids and sugars in model systems at temperatures between 125°C and 200°C. Significantly, these reactants also occur naturally in the muscle tissues of meat and fish at various concentrations. One of the most potent of the heterocyclic amines in terms of its gene-altering activity is 2-amino-3-methyl-3H-imidazo [4,5-f] quinoline. For this activity, and presumably interaction with DNA, it has first to be activated metabolically by cellular cytochrome P-450 systems, as described for a number of compounds in section 2.1.

2.5.4 PLANT PHENOLICS

Extracts of coffee beans, consumed at a high level in the Western world, are mutagenic towards certain bacteria. Although, as mentioned above, studies have indicated that roasted coffee contains at least 19 chemical compounds that cause cancer in experimental animals *when administered in high doses* (e.g. caffeic acid, chlorogenic acid, catechol, furfural, hydroquinone), the present epidemiological evidence is insufficient to identify coffee as a risk factor for cancer in humans. Tea preparations are also complex chemically, and are also mutagenic towards bacteria. Again epidemiological evidence has not so far demonstrated any strong associations of tea drinking with cancer, despite observations that some tea components can be carcinogenic in experimental animals when administered at high doses.

Both coffee and tea appear to contain hydrogen peroxide generating systems, and it is possible that the mutagenic activity of tea and coffee towards bacteria may be partly due to hydrogen peroxide. Since extracts of green coffee beans have a low capacity to generate hydrogen peroxide, the generating system seems to be associated with the roasting process and the phenolic compounds derived from the pyrolysis of phenolic amino acids, lignins, celluloses and sugars (e.g. caffeic acid, Figure 2.16, and chlorogenic acid, an ester of caffeic acid with quinic acid) may be implicated in the generation of hydrogen peroxide. It is possible that within cells the hydrogen peroxide interacts with catalytic iron in Fenton-type reactions to generate highly reactive oxidants, similar to hydroxyl radicals, which

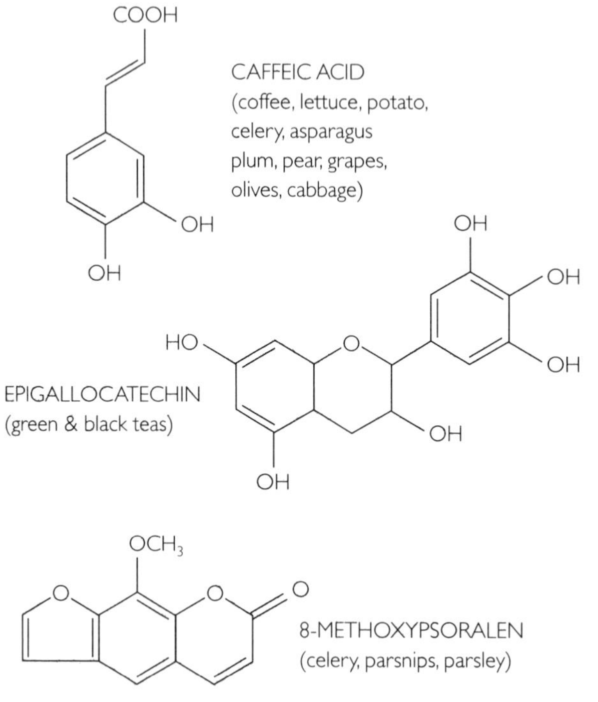

FIGURE 2.16 *Structures of caffeic acid, epigallocatechin and 8-methoxypsoralen*

can oxidatively damage DNA bases and cause strand breaks.

In tea, the processing of the leaves encourages the oxidation of abundant polyphenols such as catechins (Figure 2.16) into the polymers (tannins) which confer brewed tea with its colour and taste. A suggestion is that the hydrogen peroxide generated in tea infusions originates from the non-enzymic autoxidation of phenolics.

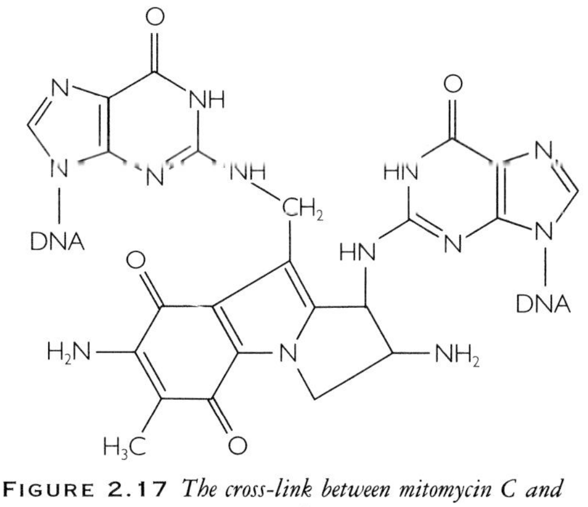

FIGURE 2.17 *The cross-link between mitomycin C and DNA strands*

2.6 *Therapeutic Drugs*

In addition to the types of toxic synthetic chemicals described in Section 2.1, it is clear that certain apparently therapeutic drugs can also induce novel types of damage to cellular DNA. Many anti-tumour drugs do not kill cancer cells immediately, instead they are designed to prevent cell division with the DNA replication phase as their main target. Since they are perhaps not strictly 'environmental' hazards, only a few examples are very briefly presented below.

2.6.1 MITOMYCIN C

Mitomycin C, used in the treatment of gastrointestinal carcinomas and bladder cancers, requires bioreductive activation before cross-linking DNA strands by covalent binding. Such interstrand cross-linking is an important type of chemical damage to DNA (Figure 2.17), as it prevents DNA strand separation and can thus completely block the vital genetic processes of DNA replication and transcription into RNA (see Chapters 3 and 7 respectively). It is believed that the covalent binding of mitomycin C to DNA occurs through a two-electron reduced form, the reduction being

achieved through the activity of a cellular enzyme, *NAD(P)H: quinone acceptor oxidoreductase (quinone reductase, DT-diaphorase)*, located in the proximity of nuclear DNA.

2.6.2 NITROGEN MUSTARDS

The use of nitrogen mustards in cancer therapy was developed after observations on the effects that mustard war gases have on cell growth. Although the sulphur war gases were too toxic for clinical use, the related nitrogen mustard alkylating agents gave rise to the earliest effective antineoplastic agents, one of which, *mechlorethamine* is still in use today. As mentioned in section 2.1, this drug has two alkylating groups that facilitate the formation of covalent cross-links between the N-7 position on guanine bases of DNA strands (Figures 2.18 and 2.2).

Further examples of clinical compounds that act by alkylation cross-linking of DNA are *cis-platinum(II)diaminodichloride*, sometimes referred to as cisplatin (see Figure 2.2), used against ovarian and germ cell neoplasms, and the antitumour diaziridinylbenzoquinones, one of which, *AZQ*, requires intracellular bioreduction before it can cross-link DNA strands.

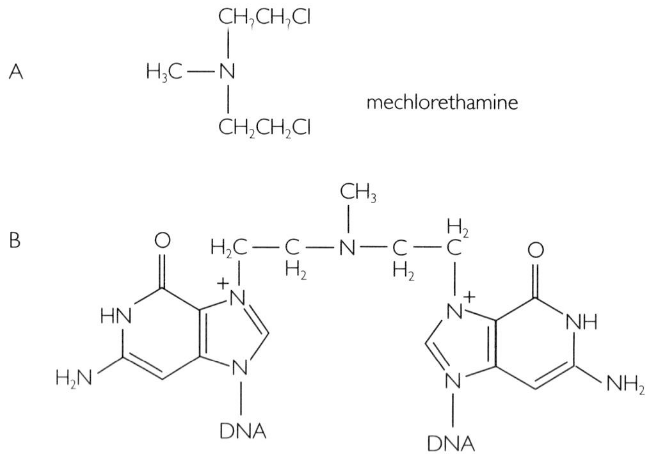

FIGURE 2.18 *Inter-strand cross-linking of DNA involving nitrogen mustard (mechlorethamine)*

2.6.3 BLEOMYCIN

Bleomycin, a mixture of several relatively high molecular weight basic glycopeptides, which is used in combination therapy for the treatment of squamous cell carcinomas, testicular carcinomas and cancers of the head and neck, appears to require the formation of an iron–bleomycin–oxygen complex for activity. The complex binds to DNA and causes the removal of bases and the creation of DNA strand breaks. The reactive species responsible for such DNA damage is unclear. Although there is strong evidence for hydroxyl radicals, it is also possible that a type of 'oxo-iron' species is involved.

2.6.4 INTERCALATING DRUGS

Anticancer drugs such as *daunorubicin* and *doxorubicin* are anthracyclines that act in part through *intercalation*, or insertion, between the stacked base pairs in double-stranded DNA (Figure 2.19). Other intercalating antitumour drugs include *actinomycin D*, but in

addition acridine dyes such as *proflavine* and *ethidium bromide* can also intercalate between DNA bases. The process of intercalation causes local unwinding of the DNA double helix and this interferes with the transcription and DNA replication processes. At replication extra DNA nucleotides can be inserted, or deleted, opposite the intercalated drug or dye.

2.6.5 PSORALENS

Psoralens are furocoumarins with planar tri-cyclic configurations (see *8-methoxypsoralen* in Figure 2.16). They can intercalate into DNA molecules, but following activation with long wavelength ultraviolet radiation they can subsequently form covalent adducts with adjacent pyrimidine bases in the DNA by addition across the 5,6 double bonds. The adducts formed can be monofunctional or difunctional. In the latter case pyrimidine bases on one DNA strand can be cross-linked to pyrimidines of the other strand through the psoralen residue. This ability to cross-link cellular DNA after photoactivation and prevent its replication has facilitated the use

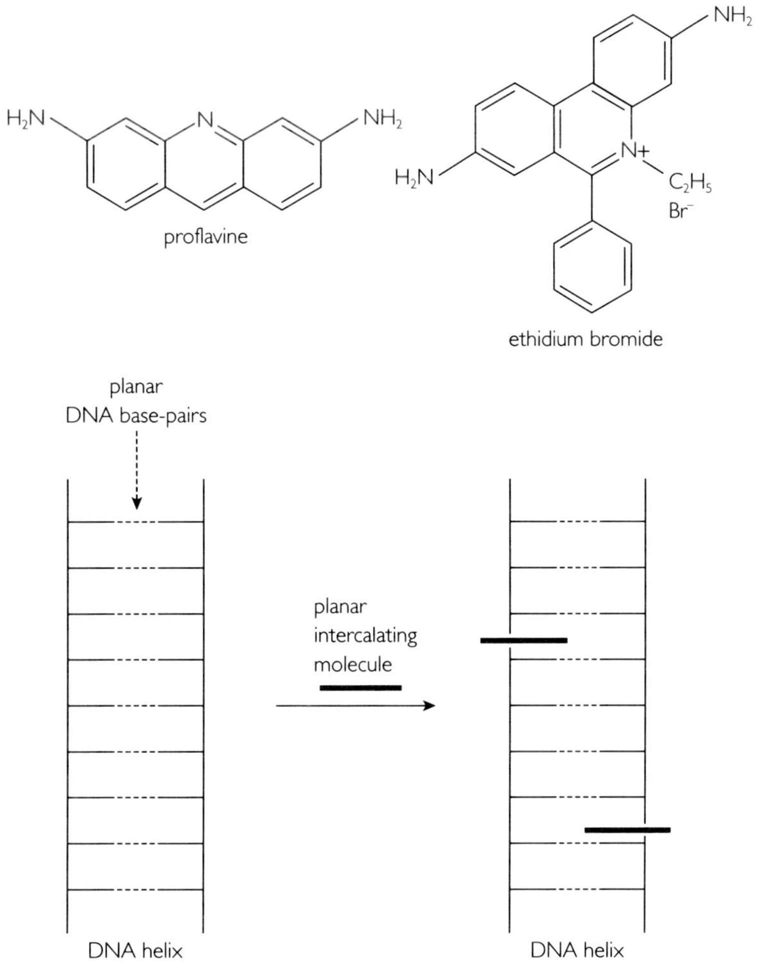

FIGURE 2.19 *A schematic illustration of the intercalation of a planar dye molecule such as proflavin or ethidium bromide between DNA base pairs*

of psoralens in the treatment of psoriasis, a condition characterized by an excessive proliferation of skin cells. Following application of the psoralens to afflicted skin tissues, it is particularly easy in dermatology clinics to carry out the necessary photoactivation step with suitably directed ultraviolet radiation.

Psoralens also occur naturally in foodstuffs. For example, 8-methoxypsoralen (Figure 2.16) is present in celery, parsnips and parsley. If taken in excessive quantities, such foodstuffs could be potentially damaging to DNA within the cells of the recipient, should these

cells become exposed to excessive ultraviolet radiation from sunlight or other sources.

2.6.6 TOPOISOMERASE INHIBITORS

Although *daunorubicin* and *doxorubicin* can act as intercalators, they can also act as inhibitors of an enzyme in the nuclei of eukaryotic cells called *topoisomerase II*. This enzyme plays a key role in catalyzing the uncoiling of DNA, for example during the complex processes of replication (see Chapter 3). This is achieved through DNA breakage and rejoining steps. Specifically, the drugs can inhibit the ability of the topoisomerase to catalyze the DNA rejoining steps, leaving broken DNA strands with the eventual death of the affected cells.

2.7 Microbial and Viral Pathogens

The previous sections of this chapter have illustrated an assortment of molecular mechanisms whereby a diverse range of environmental factors can cause damage to DNA. These took the form of damage to individual DNA bases or deoxyribose components, as well as the cross-linking, or breakage, of the individual nucleotide chains of the basic DNA helical structure. Other types of environmental problem can threaten the integrity of an organism's DNA, but in a different way. For example, certain pathogenic microbes and viruses can infect cells of both prokaryotic and eukaryotic organisms and in the process extensively alter the DNA nucleotide sequences of the host cell chromosomes. To illustrate the principle of this, specific examples are briefly presented (for more detailed accounts of this phenomenon, the reader is referred to specific texts in microbiology and virology).

2.7.1 VIRUSES

Viruses are obligate parasites that can infect eukaryotic and prokaryotic cells. They can only reproduce within a host cell. Infection begins when the genetic material (DNA or RNA) of a virus makes its way into a cell. The mechanism of entry varies depending on the type of virus. Within the cell the viral nucleic acid can commandeer its host,

reprogramming the cell to copy the viral genes and manufacture the proteins that will coat these newly synthesized genes. Most DNA-containing viruses use the DNA-replicating enzymes of the host cell to make many copies of their DNA. In contrast, RNA-containing viruses usually contain special enzymes of their own to initiate the replication of their RNA.

A group of RNA-containing viruses called the *retroviruses* has an extremely complex reproductive cycle. The group contains viruses that can induce different types of cancer in humans and animals, as well as the human immunodeficiency virus (HIV) which causes AIDS. The retroviruses carry with them a unique enzyme called reverse transcriptase which, within the infected host cell, converts the RNA of the invading virus into an equivalent molecule of DNA. This DNA equivalent then integrates *randomly* into the chromosomal DNA of the infected host cell, where it serves as template for the manufacture of further copies of viral RNA (Figure 2.20). The process thus brings about a major and permanent alteration of the deoxynucleotide sequence of the host cell chromosomal DNA at the site of integration of the 'DNA equivalent'. The host cell DNA sequence can be interrupted randomly with a short stretch of DNA of viral origin, which can have deleterious consequences for the subsequent life of the cell.

Some DNA-containing viruses are also capable of integrating their nucleic acid into the chromosomes of their hosts. For example, while the *herpesviruses* can reproduce within the nuclei of human and animal cells using a combination of viral and cellular enzymes to copy their DNA, the virus DNA itself can sometimes also become integrated into the host cell chromosome.

A similar situation exists with the well-studied bacterial virus, *bacteriophage lambda*,

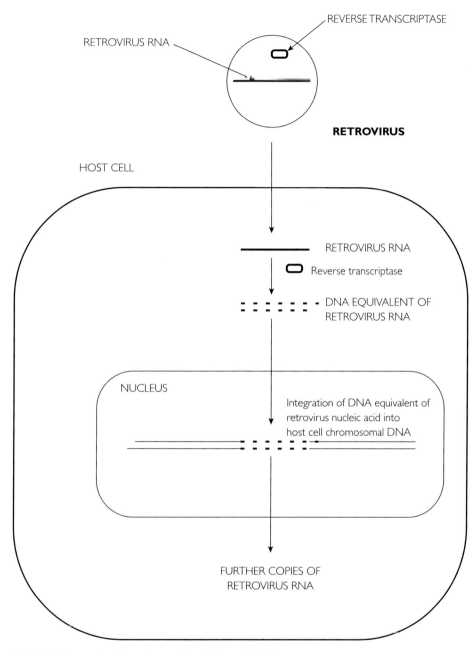

FIGURE 2.20 *The integration of retrovirus nucleic acid into the host chromosome*

which has two modes of reproduction. In one of these, the *lytic* cycle, the viral DNA reprogrammes the infected cell to become a virus-producing factory and the cell soon bursts (or lyses) to release progeny viruses. In the *lysogenic* cycle, however, the lambda bacterio-phage DNA is integrated into a specific site in the host cell chromosome. Subsequently every time the bacterium divides, it replicates the bacteriophage DNA along with its own.

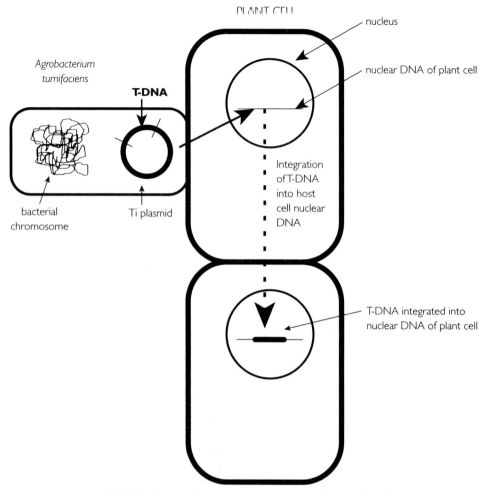

FIGURE 2.21 *Integration of T-DNA from* Agrobacterium tumifaciens *into chromosomal DNA of plants*

This type of mechanism allows the bacterio-phages to propagate without eliminating the host cells upon which they depend. Not only can bacteriophage DNA become integrated into host chromosomes, but the reverse can also occur. Sometimes an environmental trigger, such as radiation or certain chemicals, can switch the bacteriophage from the lysogenic to the lytic mode. Moreover, sometimes during the complex process of excising the bac-teriophage DNA from the host chromosome small fragments of the host chromosome may be carried along with the bacteriophage DNA.

If a bacteriophage carrying a piece of the original host DNA infects a new host, it can integrate this old host DNA into the chromosome of the new host. In this way the bacteriophage has acted as a vehicle for moving extensive stretches of DNA from the chromosome of one bacterial cell to the chromosome of another. The process is called *bacteriophage transduction.*

In some special cases, however, a vehicle is not necessary to move DNA. For instance, a limited range of bacteria, e.g. *Streptococcus*, *Bacillus* and *Haemophilus* spp., can take up DNA fragments directly from their environment, often from dead bacteria. The imported DNA fragments become integrated into their chromosomal DNA and the overall process is referred to as *transformation*.

2.7.2 MICROBIAL PATHOGENS

Most bacterial cells contain *plasmids*, which are circular DNA molecules that normally exist within bacteria and replicate independently of the main bacterial chromosome. Some plasmids are relatively small and carry only one or two genes, while others may be up to a fifth of the size of the main chromosome. Plasmids with genes that code for resistance to various antibiotics, including streptomycin, tetracycline and ampicillin, are important hazards in clinical medicine. When bacteria are exposed to such antibiotics, the number of plasmids is increased. Moreover, 'horizontal' gene transfer mechanisms exist, such as *conjugation*, whereby plasmids can be transferred from one bacterium to another, thus increasing the hazards of antibiotic resistance and bacterial pathogenicity in the clinical environment. Plasmid transfer can also occur between yeasts.

A notable microbial pathogen of plants that exerts its effects through a plasmid transfer is the free-living Gram-negative soil bacterium *Agrobacterium tumifaciens*. Virulent strains are able to infect wounded dicotyledonous (but not monocotyledonous) plants and induce the production of localized gall, or neoplastic growth, by the plant. These virulent *Agrobacterium* strains contain a very large plasmid (130 000–230 000 base pairs) known as the tumour-inducing or *Ti plasmid*. Strains of *Agrobacterium* that are non-virulent lack the Ti plasmid. It is the information encoded in the DNA of part of the Ti plasmid (T-DNA) that determines the properties of the plant gall induced following bacterial infection of the wounded plant tissue. During the process of infection the T-DNA portion of the Ti plasmid becomes transferred and integrated into the chromosomal DNA of the host plant cell (Figure 2.21). There the T-DNA segment reprogrammes the infected plant cell to synthesize unusual amino acids known as opines, which serve specifically as nutrients to support the invading bacteria. Overall the process brings about a major alteration to the plant cell genome at the particular sites of integration, which at present appear to be random. Effectively as a result of pathogen infection, the plant cell DNA nucleotide sequence is interrupted with a stretch of DNA derived from a bacterial plasmid.

2.8 Literature Sources and Additional Reading

ALLEMAN, J.E. and MOSSMAN, B.T., 1997, Asbestos revisited, *Scientific American*, July, 54–57.

ALPER, T., 1979, *Cellular Radiobiology*, Cambridge: Cambridge University Press.

AMES, B.N. and GOLD, L.S., 1997, Environmental pollution, pesticides and the prevention of cancer – misconceptions, *FASEB Journal*, 11, 1041–52.

BRASH, D.E., 1997, Sunlight and the onset of skin cancer, *Trends in Genetics*, 13, 410–14.

BRIVIDA, K., KLOTZ, L.-O. and SIES, H., 1997, Toxic and signaling effects of photochemically or chemically generated singlet oxygen in biological systems, *Biological Chemistry*, 378, 1259–65.

BUTLER, J. and HOEY, B.M., 1993, Redox cycling drugs and DNA damage, in Halliwell, B. and Aruoma, O.I (Eds) *DNA and Free Radicals*, pp. 243–73, New York: Ellis Horwood.

CADET, J., BERGER, M., DOUKI, T., MORIN, B., RAOUL, S., RAVANAT, J.-L. and SPINELLI, S., 1997, Effects of UV and visible radiation on DNA – final base damage, *Biological Chemistry*, 378, 1275–86.

CANN, A.J., 1993, *Principles of Molecular Virology*, London: Academic Press.

CUNNINGHAM, W.P. and SAIGO, B.W., 1990, *Environmental Science – a Global Concern*, Dubuque, IA: Wm C. Brown Publishers.

DIZDAROGLU, M., 1993, Chemistry of free radical damage to DNA, in Halliwell, B. and Aruoma, O.I. (Eds) *DNA and Free Radicals*, pp. 19–39, NewYork: Ellis Horwood.

FRIEDBERG, E.C., WALKER, G.C. and SIEDE, W., 1995, *DNA Repair and Mutagenesis*, Washington, DC: ASM Press.

HALLIWELL, B. and GUTTERIDGE, J.M.C., 1998, *Free Radicals in Biology and Medicine*, 3rd edn, Oxford: Clarendon Press.

HENLE, E.S. and LINN, S., 1997, Formation, prevention and repair of DNA damage by iron/hydrogen peroxide, *Journal of Biological Chemistry*, 272, 19095–98.

HUGHES, M.A., 1996, *Plant Molecular Genetics*, Harlow: Addison Wesley Longman.

IZZOLTI, A., BALONSKI, R.M., BLAGOEVA, P.M., MIRCHEVA, Z.I., TULIMIERO, L., CARTIGLIA, C. and DeFLORA, S., 1998, DNA alterations in rat organs after chronic exposure to cigarette smoke and/or ethanol ingestion, *FASEB Journal*, 12, 732–35.

KELLY, F.J., 1996, Air pollution: an old problem in a new guise, *Biologist*, 43, 102–5.

LAWRENCE, C.W., 1979, *Cellular Radiobiology*, London: Edward Arnold.

LEANDERSON, P. and TAGESSON, C., 1993, Mineral fibers, cigarette smoke and oxidative stress, in Halliwell, B. and Aruoma, O.I. (Eds) *DNA and Free Radicals*, pp. 293–314, New York: Ellis Horwood.

PUEYO, C. & ARIZA, R.R., 1993, Role of reactive oxygen species in the mutagenicity of complex mixtures of plant origin, in Halliwell, B. and Aruoma, O.I. (Eds) *DNA and Free Radicals*, pp. 276–91, New York: Ellis Horwood.

RICE-EVANS, C.A., 1994, Formation of free radicals and the mechanisms of action in normal biochemical processes and pathological states, in Rice-Evans, C.A. and Burdon, R.H (Eds) *Free Radical Damage and its Control*, pp. 131–154, Amsterdam: Elsevier Science.

VENITT, S. and PHILLIPS, D.H., 1995, The importance of environmental mutagens in human carcinogenesis and germline mutation, in Phillips, D.H. and Venitt, S. (Eds) *Environmental Mutagenesis*, pp. 1–20, Oxford: Bios Scientific Publishers.

WELLBURN, A., 1994, *Air Pollution and Climate Change, the Biological Impact*, 2nd edn, Harlow: Longman Scientific and Technical.

WELLINGTON, E.M.H. and VAN ELSA, J.D., 1992, *Genetic Interactions among Microorganisms in the Natural Environment*, Oxford: Pergamon Press.

WILLIAMS, G.M., 1993, Food: its role in the etiology of cancer, in Waldron, K.W., Johnson, I.T. and Fenwick, G.R. (Eds) *Food and Cancer Prevention: Chemical and Biological Effects*, pp. 1–11, London: Royal Society of Chemistry.

CHAPTER **3**

Spontaneous Alterations to DNA

KEY POINTS

■ The spontaneous modification of cellular DNA can be the outcome of the intrinsic chemical properties such as tautomerization, or instability, of different structural components of DNA such as individual bases or nucleotides.

■ Alterations to the DNA of an organism can also result from errors that can occur at low frequency, despite proof-reading mechanisms, during the complex enzymic processes associated with the replication of cellular DNA.

■ The chapter also broadly summarizes current understanding of the basic biochemical processes involved in the replication of cellular DNA and the mechanisms available within cells to eliminate possible errors.

Many types of environmental factor are hazardous to genes because they can bring about damage and alteration to cellular DNA. However, such damage is superimposed on a considerable amount of DNA change that already takes place spontaneously and continuously within organisms. In this chapter the basic enzymic mechanisms involved in the replication of cellular DNA are briefly reviewed. Primarily the aim is to provide a basis for the appreciation of the intrinsic chemical instability of DNA itself, as well as the mechanisms available to organisms to minimize the occurrence of errors during the replication process and thus the frequency of mutation. Additionally, as will be apparent in subsequent chapters, many components of the DNA replication apparatus described also play crucial roles in the enzymic repair of damaged DNA.

3.1 *Gene Replication and 'Proof-reading' Mechanisms for Error Correction*

Besides encoding genetic information, genes serve as a pattern for the manufacture of further copies of themselves. When cells divide and sperm and eggs are produced, a team of enzymes and other proteins functions to replicate the genes, so that every new cell has a complete set of genes more or less identical to the genes in the original cells. The chemical processes by which the nucleotide sequence of DNA of the genes is duplicated into two copies are remarkable for their almost perfect fidelity. The majority of the few mistakes that occur during the replication process are corrected by repair mechanisms that scan the DNA to detect nucleotide mismatches and

other irregularities. DNA and RNA are unique in their ability to direct their own replication.

When Watson and Crick elucidated the double-stranded complementary base-paired helical structure of DNA in 1953, they pointed out that a possible replication mechanism was inherent in its structure. Because the two strands of a DNA molecule are 'complementary' in terms of their nucleotide sequence, each can serve as a template for the formation onto itself of a new complementary companion strand when the individual nucleotide strands of the helical DNA molecule separate by unwinding (Figure 3.1). Eventually two pairs of nucleotide strands will arise by this *semi-conservative* process which, although simple in outline, is complex in its biochemical detail.

More than a dozen enzymes and other proteins are now known to participate in the overall DNA replication process, and although the bulk of the detailed mechanistic information has been obtained from bacterial systems, most of the basic steps in replication seem to be similar for prokaryotes and eukaryotes. These are briefly summarized in the following sections.

3.1.1 MAIN FEATURES OF DNA REPLICATION

Copying DNA

The assembly of individual nucleotides into new DNA strand is catalyzed by a group of enzymes known as *DNA polymerases*. These all use the four DNA nucleotides as building blocks but in the activated form of deoxynucleoside 5'-triphosphates, dATP (deoxyadenosine triphosphate), dGTP (deoxyguanosine triphosphate, dCTP (deoxycytidine triphosphate, and dTTP (deoxythymidine triphosphate). A key characteristic of DNA polymerases is

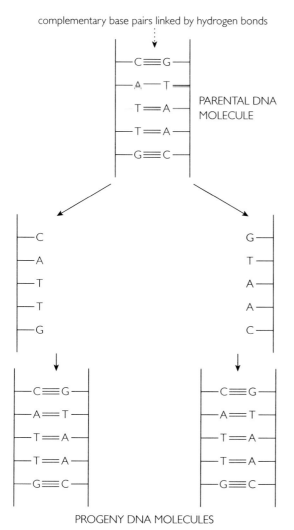

complementary base pairs linked by hydrogen bonds

PARENTAL DNA MOLECULE

PROGENY DNA MOLECULES

FIGURE 3.1 *Schematic illustration of the separation of the two complementary strands of a portion of DNA with formation of a new complementary strand on each as the outcome of a semi-conservative process: interstrand complementary base pairs between adenine (A) and thymine (T) as well as guanine (G) and cytosine (C) are indicated, linked with 2 and 3 hydrogen bonds respectively*

that they can initiate synthesis only by adding nucleotides to the ends of existing nucleotide chains that therefore act as 'primers'.

As shown in Figure 3.2, polymerization of nucleotide building blocks proceeds from DNA. Initially DNA polymerase binds to

the 3'-OH group of the primer and recognizes the nature of the first nucleotide base to be copied from the 'template' strand. If this is adenine then by the 'base-pairing' rules the polymerase will bind dTTP from the deoxynucleoside triphosphates in the surrounding medium. While all four deoxynucleoside triphosphates can bind weakly to DNA polymerases, normally only the deoxynucleotide forming the correct complementary base pair with the template nucleotide base (dTTP in this case) will proceed from the initial weak binding to the tight binding. This then specifically holds the 'complementary' nucleotide triphosphate dTTP in a such a position that favours the establishment of a covalent bond between the terminal 3'-OH group of the primer and the innermost phosphate group of the dTTP, and the splitting off of the outer two phosphates of the dTTP. Overall this process creates a 3'–5' phosphodiester linkage between the primer and the first added nucleotide (Figure 3.2). The DNA polymerase then processes to the next nucleotide base in the template strand. If this is cytosine the polymerase will tightly bind a 'complementary' dGTP molecule and promote the formation of the second 3'–5' phosphodiesterase link in a fashion analogous to that already described. This process is repeated, 'complementary' nucleotides being added in succession to the strand of new DNA growing from the primer. Each new nucleotide added to an available 3'-OH then provides the next 3'-OH group for the next nucleotide addition. A 3'-OH group is always to be found at the 'new' end of a growing strand. For this reason the synthesis of new DNA strands is said to proceed in a 5' to 3' direction and all known DNA polymerases work in this direction. The two nucleotide strands of a DNA molecule, however, have opposite 'polarity', i.e. they run in opposite directions when their 3'–5' phosphodiester linkages are

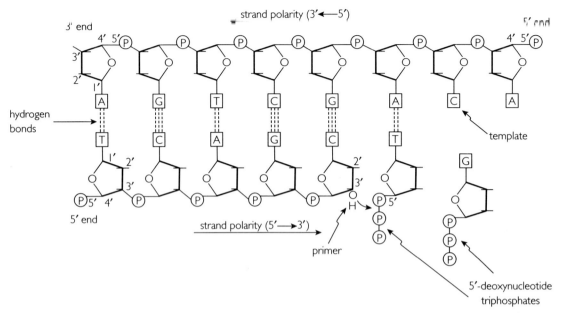

FIGURE 3.2 *A representation of DNA structure illustrating the mode of action of DNA polymerase using one strand of DNA as template and the other as primer*

examined. Thus the newly synthesized DNA in Figure 3.2 runs in a direction opposite to the 'template' strand, in chemical terms.

Unwinding DNA

During the replication of DNA within cells it is first necessary for the DNA double helix to be unwound. This is achieved by a special type of enzymatic protein called a *DNA helicase*. This protein appears to move along the double helical molecule ahead of the DNA polymerase, separating the component template strands. Once unwound, the template nucleotide strands are stabilized by *single-stranded binding proteins* which, as their name implies, bind avidly to the separated single strands to prevent them from rewinding (Figure 3.3). This binding appears to be co-operative such that several molecules bind in rows to the unwinding strands.

However, as unwinding of the DNA helix proceeds, a problem builds up in front of the helicase because of the right-handed direction of the DNA helix turns. The helical DNA in front of the helicase cannot rotate readily to allow for the twisting. Rather, the DNA in front is forced into one 'supercoil' for each turn unwound by the helicase. Another group of enzymes, the *DNA topoisomerases*, have the ability to solve the problem of supercoils. They can break the DNA ahead of the helicase and allow the DNA to rotate around the site of the break and then reseal the DNA (Figure 3.3). There are two types of topoisomerase: *DNA topoisomerase I and II*. In the case of *DNA topoisomerase I*, a cut is introduced into one of the two DNA strands. The enzyme then permits the uncut strand to pass through the opening created in the cut strand and then reseals that cut. *DNA topoisomerase II*, which may be more important in the cell (and is also known as *DNA gyrase*), in contrast breaks

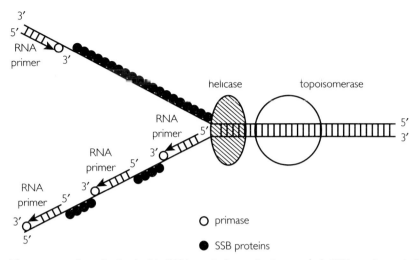

FIGURE 3.3 *The enzymes and proteins involved in DNA unwinding and primer synthesis (SSB proteins = single-stranded binding proteins)*

both strands and allows a portion of the intact DNA double helix to pass through the break point before resealing it, thus relieving the supercoil. To achieve this, the enzyme requires energy in the form of ATP (or GTP).

Priming the replication process

Although both DNA and RNA can act as primers for DNA synthesis, RNA is initially used as a primer for the process of DNA replication within living cells. Following the unwinding processes, RNA primers are created through the activity of an enzyme called a *primase*, which, in bacteria at least, is associated with the helicase described above. In reality this is a special type of RNA polymerase which binds to the unwound template strands and catalyzes the assembly, in the 5′ to 3′ direction, of small RNA primers (5–10 nucleotides). Such primer synthesis appears to displace the single-strand binding proteins. (In eukaryotes, it should be noted that the primase activity is associated not with the helicase but rather with DNA polymerase

α, helicase activity being a feature of the eukaryotic DNA polymerase δ, as detailed below.) Deoxyribonucleotides are then added sequentially to the ends of the RNA primers, synthesis of new DNA proceeding in the 5′ to 3′ direction.

Once the primers have initiated the synthesis of new DNA strands, the RNA primers can be removed by the DNA polymerases themselves. This appears to depend on the ability of certain DNA polymerases to function not only as a polymerizing enzyme in the 5′ to 3′ direction, but also as a *5′ to 3′ exonuclease*. In this latter role the polymerase removes RNA nucleotides one at a time from the primer. As it removes the RNA nucleotides it adds deoxyribonucleotides in their place (Figure 3.4).

The replication fork

As can be seen from Figure 3.3, when the two strands of a DNA molecule are unwound a 'fork-like' structure is evident. Initial studies suggested that DNA replication proceeded unidirectionally as the DNA unwound at

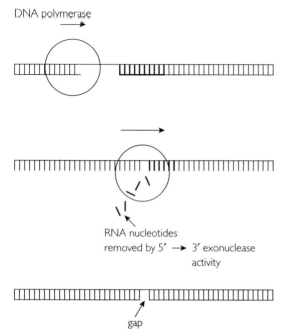

FIGURE 3.4 *Removal of RNA primer (in bold) using the 5′ to 3′ exonuclease activity of DNA polymerase*

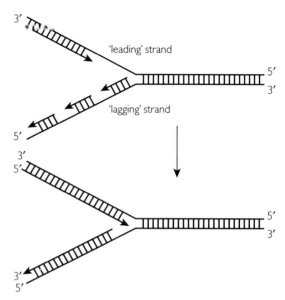

FIGURE 3.5 *While the 'leading' strand acts as template for continuous DNA synthesis, the 'lagging' strand is also used as template, but in short bursts and in the opposite direction; these short lengths are then linked into a continuous chain, through the action of DNA ligase*

this fork. However, this presents a physico-chemical difficulty as the two nucleotide strands of the unwound DNA would have to be replicated in the same direction whereas, as already pointed out, the two chains of a DNA molecule actually run in opposite directions. In the cell this problem is solved by a mechanism whereby the replication of the 'wrong way' strand is actually achieved by short bursts of DNA synthesis that occur in the direction opposite to the general fork movement. The DNA strand that is replicated in the direction of the fork movement is generally referred to as the 'leading' strand, which is replicated in a continuous manner.

The strand copied in the opposite direction is called the 'lagging' strand (Figure 3.5). RNA primers are generally laid down on the lagging strand at intervals of 1000 to 2000 nucleotides in bacteria and 100 to 200 in eukaryotes. The discontinuous lengths of

new DNA that are made by extension from these primers by DNA polymerases (sometimes referred to as *Okazaki fragments*) are subsequently joined together, after removal of the RNA primers, into a continuous strand of new DNA (Figure 3.5). This is achieved through the catalytic activity of specialized joining enzymes called *DNA ligases.* The combination of all these enzymic activities results in DNA replication which in an overall sense proceeds in the direction of the fork movement at a rate of around 50 to 100 nucleotides per second in eukaryotes and some ten times that rate in bacteria.

3.1.2 THE DNA POLYMERASES

In the bacterium *Escherichia coli* three distinct DNA polymerases have been identified. These are designated I, II and III. All three can act to polymerize deoxynucleotides in the 5′ to 3′ direction. They also act as DNA exonucleases. While they all possess 3′ to 5′ exonuclease activity, polymerases I and III also manifest 5′ to 3′ exonuclease activity. As a result of genetic analysis of temperature-sensitive mutants, it was established that DNA polymerase III was the primary replication enzyme in *E. coli*, although DNA polymerase II may play an auxiliary role. The role of polymerase I in the replication process seems relatively minor, possibly filling in gaps left after removal of primers. It appears to play a more significant role in the mechanisms that contribute to the 'repair' of damaged DNA (see Chapter 5).

DNA polymerase III is a very complicated protein comprising some 20 polypeptide subunits. However the 'core' subunit occurs in two copies and is responsible for the basic polymerization activity. Two copies are also present of subunit carrying the 3′ to 5′ exonuclease activity. By comparison, DNA polymerase I is relatively simple, consisting of only a single polypeptide chain but with both polymerizing and exonuclease activity.

Studies on higher eukaryotes has also indicated a multiplicity of DNA polymerases, but these number five rather than three. All five, namely α, β, γ, δ and ε, are like their bacterial counterparts inasmuch as they can only catalyze the synthesis of new DNA strands in the 5′ to 3′ direction. It seems, however, that the primary replication enzyme is DNA polymerase α. Although this enzyme lacks any exonuclease activity, two of its subunits have RNA primase activity. Additionally, DNA polymerase δ appears to be closely involved in eukaryotic DNA replication. Polymerase δ does have 3′ to 5′ exonuclease activity and can also act as a helicase (see above). It is believed that polymerase δ is most important in catalyzing the synthesis of the 'leading' strand in eukaryotic DNA replication and

polymerase α the lagging strand, the latter enzyme also having RNA primase activity.

DNA polymerase γ is found predominantly in mitochondria and for this reason is thought to have a special role in the replication of mitochondrial DNA. As will be discussed in Chapter 5, DNA polymerase ε has been implicated in the repair of damaged DNA molecules, as has DNA polymerase β.

3.1.3 EXONUCLEASES AND THE CORRECTION OF ERRORS DURING NORMAL DNA REPLICATION: 'PROOF-READING'

The role of the exonuclease activities associated with the DNA polymerases has been a subject of much research. It appears, for example, that the 3′ to 5′ exonuclease activity of DNA polymerase III of *E. coli* plays a key role in the correction of any base-pairing 'errors' that occur during the copying of template strands. Apparently DNA polymerase III will only add nucleotides to a new strand growing from a primer if the most recently added nucleotide is correctly base-paired. Should this be incorrectly paired and thus unstable, the 3′ to 5′ exonuclease activity of DNA polymerase III is favoured and the enzyme reverses, removes the incorrect nucleotide before proceeding with its normal polymerization activities (Figure 3.6). Overall the outcome of this polymerization combined with exonuclease activity is to effect a 'proof-reading' of the new progeny DNA strand under construction, eliminating any base-pairing mistakes as soon as they happen. While the overall error rate as a result of this combined activity is around 10^{-6}, an important consideration is that if the exonuclease activity is blocked, then the error rate can rise to 10^{-4} or even 10^{-3}. In the case of the eukaryotic DNA polymerases, present evidence

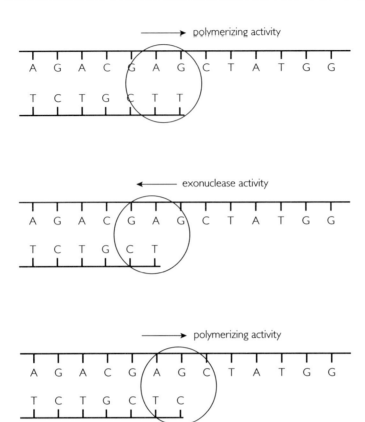

FIGURE 3.6 *'Proof-reading' by DNA polymerase*

suggests that the exonuclease activity of polymerase δ provides the 'proof-reading' function.

3.2 *Errors during DNA Replication Arising from Base Tautomerization*

While DNA polymerases are highly accurate for the above reasons and ultimately enter the appropriate complementary deoxynucleotide opposite the nucleotide in the template strand with a high frequency, there are other causes of low-frequency base mismatch. One natural source arises from the rare occurrence of *tautomeric*, or alternative, configurations of the four

normal DNA nucleotides that allow abnormal nucleotide pairs to form in DNA. Around one in every 10 000 to 100 000 nucleotide bases can undergo a change to a different tautomeric form at any time. While these forms may persist for only a short time, they can lead to a nucleotide 'mismatch' if they occur, for example, in a DNA strand that is about to be used as a template in the replication process. For instance, a change in guanine from its predominant 'keto' form, which normally pairs with cytosine, to the tautomeric 'enol' form allows it to base-pair with thymine instead. Similarly, a tautomeric form of adenine ('imino' rather than 'amino' form) can erroneously pair with a cytosine nucleotide. At the following round of DNA replication, the incorrect nucleotide (containing cytosine) is

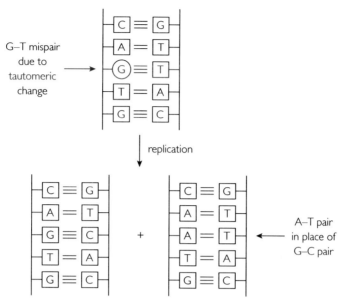

FIGURE 3.7 *An error in DNA occurring as a result of guanine base tautomerization*

inserted in the progeny strand and as a result of the tautomeric change in the template strand, adenine is copied such that the ensuing progeny strand has a thymine at that particular position. All future progeny cells then receive similarly altered DNA molecules. In this way a guanine–cytosine base pair in a DNA molecule of a cell will be supplanted by an adenine–thymine base pair in the DNA of future cell generations, and an adenine–thymine can give rise to a guanine–cytosine base-pair (Figure 3.7).

3.3 *Errors due to Deamination of DNA Bases*

A slow loss of amino groups from the bases adenine, guanine and cytosine normally found in DNA (as well as the 'minor' base 5-methylcytosine) can occur spontaneously in chemical reactions which are affected by both temperature and pH. The resultant 'new bases' from such deamination processes are hypoxanthine, xanthine, uracil and thymine respectively (Figures 3.8 and 3.9). The rate of spontaneous deamination of cytosine in single-stranded DNA, at 37°C and physiological pH, has been estimated to be equivalent to a half-life of an individual residue of about 200 years, while the corresponding time for a residue in the double-stranded form of DNA is around 100 times longer. In the cell, although it is clear that DNA exists predominantly in a double-stranded form, there are nevertheless times during the replication processes when single-stranded regions, which could potentially permit the faster rate of cytosine deamination, exist transiently. The rates of deamination of adenine and guanine under physiological conditions are around 50 times lower than the conversion of cytosine to uracil.

The deamination of cytosine residues in DNA to yield uracil residues can also be brought about by exposure to *nitrous acid* or by *sodium bisulphite*. In the case of nitrous

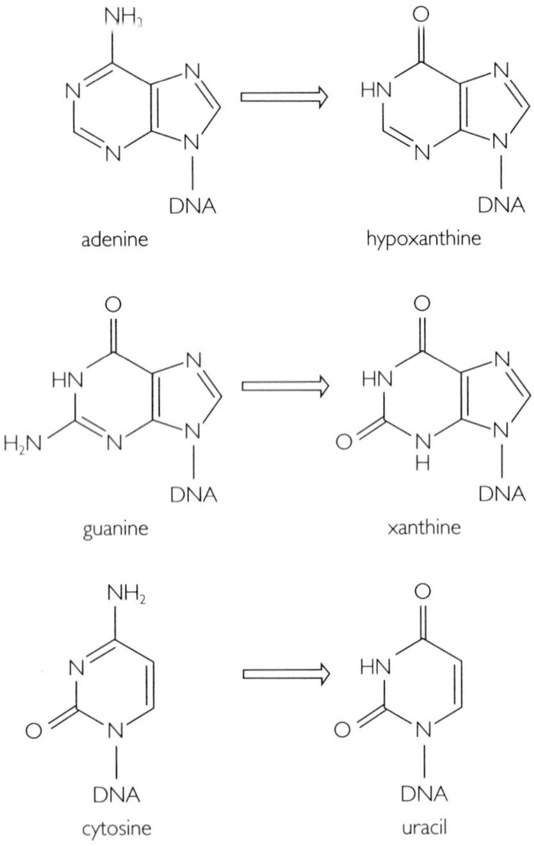

FIGURE 3.8 *Products of DNA base deamination*

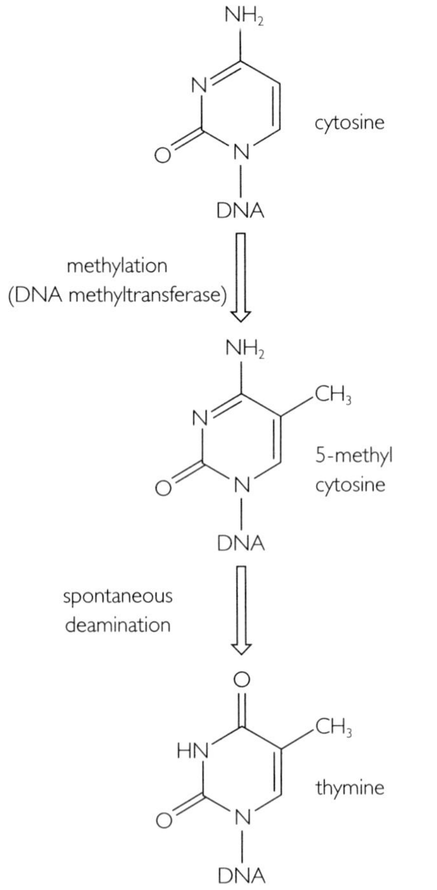

FIGURE 3.9 *The methylation and deamination of cytosine in DNA to create a thymine residue*

acid, the deamination of cytosine residues occurs with almost equal efficiency in double- or single-stranded DNA, whereas sodium bisulphite deaminates cytosine residues only within single-stranded regions of DNA.

The new bases in DNA that arise from these deamination reactions are of considerable significance from the point of view of the structure of a gene and its inheritance. For example, hypoxanthine, present in the place of an original adenine residue, at subsequent rounds of DNA replication will specify a cytosine residue in new DNA strands rather than thymine (Figures 3.10 and 3.11), thus creating a permanent and inheritable alteration in the nucleotide sequence of a gene.

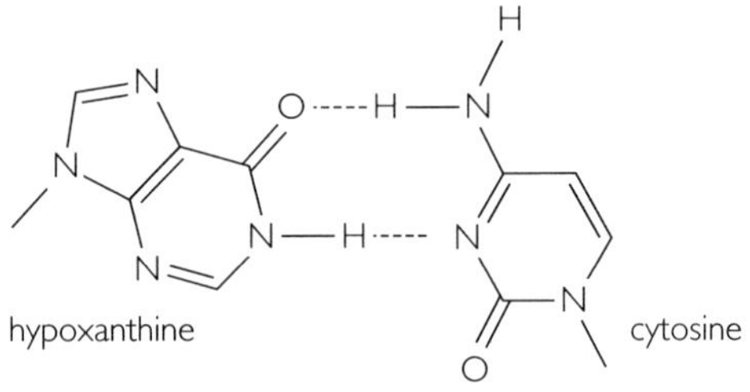

FIGURE 3.10 *Hypoxanthine–cytosine base pair*

In summary, if there were an adenine–thymine base-pair in a particular gene, there would subsequently be a guanine–cytosine base-pair. In a similar vein, any uracil arising

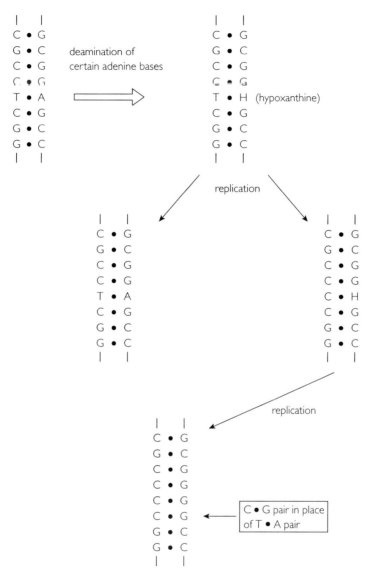

FIGURE 3.11 *The effects of adenine deamination at subsequent rounds of DNA replication (the hydrogen bonding between complementary bases is represented by •)*

from the deamination of cytosine would specify an adenine rather than a guanine residue in the progeny DNA strand at the next round of replication. This will ultimately result in an adenine–thymine base pair in place of the original guanine–cytosine. On the other hand, xanthine, arising from guanine, cannot pair in a stable fashion with either thymine or cytosine, and is believed to prevent subsequent replication of DNA.

As already mentioned, another DNA base that can undergo slow spontaneous deamination is the 'minor base' 5-methylcytosine. Unlike the four major DNA bases, 5-methylcytosine occurs only at low levels in most cellular DNAs and is formed as a result of the

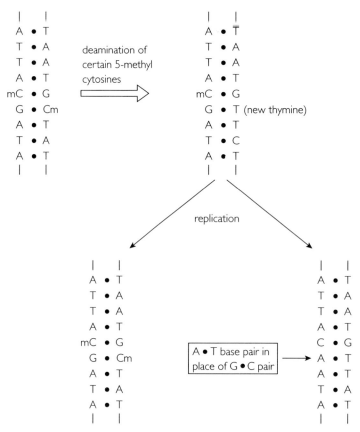

FIGURE 3.12 *The effects of 5-methylcytosine at subsequent rounds of DNA replication (the hydrogen bonding between base pairs is represented by •)*

enzymatic methylation of cytosine residues already present in cellular DNA through the action of *DNA methyltransferases*. While this modified base is believed to play biological roles in the regulation of gene expression and in gene 'imprinting', its spontaneous deamination gives rise to an apparently normal DNA base, thymine. At first sight this seems innocuous, but the thymine now present in the place of the original 5-methylcytosine will then be aberrantly base-paired with the original guanine residue in the opposite DNA strand. Following subsequent rounds of DNA replication, this situation will lead to the formation of progeny DNA molecules in which a specific cytosine–guanine base pair is altered to an adenine–thymine base pair. Thus this type of mechanism can also lead to an inheritable alteration in the nucleotide sequence of DNA (Figure 3.12).

3.4 *The Spontaneous Loss of Bases from DNA*

Spontaneous depurination (loss of purine bases, adenine or guanine from intact DNA) and depyrimidation (loss of pyrimidine bases, cytosine or thymine) has been extensively studied *in vitro* under various conditions. The

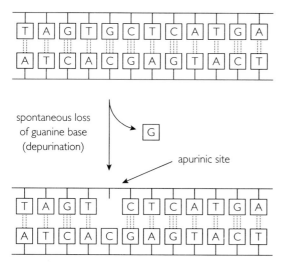

FIGURE 3.13 *Creation of an apurinic site due to spontaneous loss of a guanine base (depurination)*

mechanisms that lead to this type of DNA damage (Figure 3.13) appear to involve base protonation followed by cleavage of the glycosyl bonds between the bases and deoxribose units. From work at physiological pHs and temperatures it has been estimated that for bacteria *in vivo* the loss of purines might be as high as one residue per generation, and at higher temperatures the rate could increase considerably. In the case of mammalian cells, a spontaneous loss of 10 000 purines per generation has been suggested. In contrast, pyrimidines are generally lost at a lower rate (20%), the glycosylic bond between pyrimidines and deoxyribose units in DNA being stronger than is the case for purines and deoxyribose.

In principle, loss of DNA bases by depurination or depyrimidation would cause considerable gene damage because, during subsequent DNA replication processes, the copying machinery would not be able to specify a base complementary to the vacant apurinic or apyrimidinic site. However, as will be seen in Chapter 5, systems exist within cells to repair such sites and thereby minimize gene damage.

3.5 Literature Sources and Additional Reading

ALBERTS, B., BRAY, D., LEWIS, J., RAFF, M., ROBERTS, K. and WATSON, J.D., 1994, *Molecular Biology of the Cell*, New York: Garland Publishing.

BOSOVER, S.R., HYAMS, J.S., JONES, S., SHEPHERD, E.A. and WHITE, H.A., 1997, *From Genes to Cells*, New York: Wiley-Liss.

FRIEDBERG, E.C., WALKER, G.C. and SIEDE, W., 1995, *DNA Repair and Mutagenesis*, Washington DC,: ASM Press.

KARP, G., 1996, *Cell and Molecular Biology – Concepts and Experiments*, New York: John Wiley & Sons.

LINDAHL, T., 1993, Instability and decay of the primary structure of DNA, *Nature*, **362**, 706–15.

LODISH, H., BALTIMORE, D., BERK, A., ZIPURSKY, S.L., MATSUDAIRA, P. and DARNELL, J., 1995, *Molecular Cell Biology*, 3rd edn, New York: Scientific American Books, W.H. Freeman.

CHAPTER 4

Possible Oxidative DNA Damage from Cellular Free Radicals

KEY POINTS

- Certain types of free radical and related species are generated normally in significant amounts by both animal and plant cells.

- Although some of these free radical species serve vital cellular functions, some may also contribute, directly or indirectly, to an important background level of oxidative damage to DNA.

In Chapter 2 it was shown that components of cellular DNA could be damaged by ionizing radiation as a result of the action of hydroxyl free radicals generated from cell and tissue water molecules. It is now clear, however, that certain types of free radical, and related species, are actually generated quite normally in highly significant amounts by plant and animal cells. The aim of this chapter is to illustrate the oxidative damage that may occur in DNA as a result of cellularly generated free radicals. Although aerobic organisms gain a significant energetic advantage using molecular oxygen as the terminal oxidant in respiration, oxygen has the potential to be partially reduced in metabolic processes to yield *reactive oxygen species (ROS)*. These include singlet oxygen, superoxide radicals, hydrogen peroxide and hydroxyl radicals. Although some serve vital cellular functions, for instance in immunological protection, they are potentially hazardous in that they may contribute directly or indirectly to a low background level of oxidative damage to cellular macromolecules, including DNA.

4.1 Neutrophils, Macrophages and the Oxidative Burst

The phagocytic neutrophils and macrophages are specialized cells of the human immune system that circulate in the bloodstream seeking out foreign organisms such as bacteria and viruses, which they engulf and kill within a phagocytic plasma membrane vesicle. Part of this killing mechanism involves the activation of processes called the *oxidative burst* in which the free radical superoxide is produced. The production of superoxide is achieved from molecular oxygen through the action of a plasma membrane *NADPH–oxidase* complex.

The electrons released from the oxidation of NADPH reduce oxygen to the *superoxide radical*

$$2O_2 + NADPH \xrightarrow{\text{NADPH-oxidase}} NADP^+ + H^+ + 2O_2^{\cdot-}$$
$$\text{superoxide}$$

The NADPH–oxidase complex comprises many different subunits. In the non-stimulated cell some of these subunits are found in the cytosol while others are found in various membranes of secretory vesicles and intracellular granules. When the phagocytic cell is activated in repose to an infectious agent, all these components are transferred to the plasma membrane, where they assemble to form the active oxidase.

Following uptake of foreign particles into the plasma membrane vesicle, they are exposed to a high flux of superoxide radicals (some of which are also released extracellularly). However, superoxide itself is unlikely to be the primary killing agent in view of its comparatively low reactivity, although some bacteria can be killed by hydrogen peroxide which is formed by the spontaneous dismutation of superoxide radicals

$$2O_2^{\cdot-} + 2H^+ \xrightarrow{\text{spontaneous dismutation}} H_2O_2 + O_2$$
$$\text{hydrogen}$$
$$\text{peroxide}$$

However, once the membrane vacuole is formed, fusion with other granules in the phagocytic cell cytoplasm releases an enzyme called *myeloperoxidase*, which can use that hydrogen peroxide as a substrate, and oxidizes chloride to hypochlorous acid. This product is highly reactive, can oxidize many biomolecules and will bring about the killing of the engulfed organisms

$$H_2O_2 + Cl^- \xrightarrow{\text{myelperoxidase}} OH^- + HOCl$$
$$\text{hypochlorous}$$
$$\text{acid}$$

The primary generation of superoxide through the action of NADPH–oxidase is clearly very important. Patients with *chronic granulomatous disease* have a deficiency in this enzyme and manifest persistent multiple infections of the skin, lungs, liver and bone, due to the presence of bacteria that cannot be killed by this mechanism. Despite its vital role, this bodily defence mechanism can be hazardous. For instance, diseases such as rheumatoid arthritis and inflammatory bowel disease are accompanied by inappropriate phagocytic activation and concomitant damage to associated tissues to which these released reactive oxygen species may contribute. Indeed, recent experiments indicate that reactive oxygen species emanating from stimulated neutrophils can cause damage to the DNA of other cells in their vicinity. Recent work suggests that plants also manifest a type of 'oxidative burst' relating to mechanisms of resistance to pathogen attack. Rapid generation of superoxide and accumulation of hydrogen peroxide is a characteristic feature of the hypersensitive response following the perception of certain plant pathogen signals. A plasmalemma-based *NADPH-oxidase* is believed to be involved. This is based on evidence of cross-reactivity of antibodies raised to a component of the mammalian NADPH–oxidase and the ability to inhibit the pathogen-induced oxidative burst in certain plant cell types using diphenylene iodonium, the suicide inhibitor of the mammalian NADPH-oxidase.

4.2 *Superoxide Generation and Release from Non-phagocytic Mammalian Cells*

The release of superoxide and its dismutation product hydrogen peroxide has also been detected from non-phagocytic mammalian cells such as endothelial cells, and fibroblasts. From the use of enzyme inhibitors it appears that the generation of such species involves the participation of enzymes with many properties similar to the plasma membrane NADPH–oxidases of phagocytic cells. Although the rate of superoxide generation is low compared with that encountered in stimulated neutrophils, in most cases the rate of superoxide (or hydrogen peroxide) release is greatly stimulated when the particular cells are exposed to specific cytokines or polypeptide growth factors and/or phorbol esters (see Table 4.1). Perhaps significantly, the release does not require

TABLE 4.1 *Cellular release of superoxide radicals or hydrogen peroxide*

Cell type	Stimulus
Human fibroblasts	Cytokines/phorbol ester
Human endothelial cells	Cytokines/phorbol ester
Human fat cells	Cytokines/basic fibroblast growth factor
Human colonic epithelial cells	Deoxycholate/phorbol ester
Balb/3T3 cells	Platelet-derived growth factor/phorbol ester
Rat pancreatic islet cells	Calcium ionophore/phorbol ester
Murine keratinocytes	Phorbol ester
Rabbit chondrocytes	Cytokines/basic fibroblast growth factor
Human tumour cells	None required

cytokine or growth factor stimulation in the case of tumour cells.

While activation of plasma membrane NADPH–oxidases is likely to be affected *in vivo* by a range of ligands of environmental significance, studies are still at an early stage. However, an intriguing recent study has indicated that the generation of cellular reactive oxygen species within mammalian cells through NADPH-oxidase-like systems can be stimulated by particles of α-radiation.

4.3 *Other Cellular Sources of Superoxide or Hydrogen Peroxide*

4.3.1 MITOCHONDRIA

The oxygen molecules in air play an important role in picking up electrons when food is oxidized for energy. In particular they pick up four electrons together with four protons to make water in our mitochondria.

$$O_2 + 4e^- + 4H^+ \longrightarrow 2H_2O$$

This is a part of the normal processes underlying mitochondrial electron transport. The role of cytochrome oxidase is to keep all partially reduced oxygen intermediates tightly bound to its active centre. However, some components of the electron transport chain, by accidents of their chemistry, can leak electrons directly to oxygen. Because oxygen is able to accept one electron at a time, superoxide radicals can result

$$O_2 + e^- \longrightarrow O_2^{\cdot-} \text{ (superoxide)}$$

The superoxide free radicals are generated towards the matrix surface of the inner mitochondrial membrane and, being charged molecules, tend to stay within the inner mitochondrial compartment where they are converted to hydrogen peroxide through the enzymatic activity of *Mn-superoxide dismutase*, located within the mitochondria (see Chapter 5). Hydrogen peroxide, in contrast, can diffuse through the mitochondrial and other cellular membranes and become more distributed throughout cells.

Recent studies indicate that as much as 2–5% of the oxygen consumed by mitochondria may be converted to superoxide by this route, the overall extent being dependent on the basal metabolic rate and/or caloric utilization of particular cells.

4.3.2 CHLOROPLASTS

In plants, besides mitochondria, chloroplasts are also a source of superoxide. This is generated by the auto-oxidation of a thylakoid membrane-bound primary electron acceptor of photosystem I. While the production rate is high ($2.4 \times 10^{-4} \text{ M s}^{-1}$), the steady state level of superoxide in chloroplasts is in the region of $3 \times 10^{-9} \text{ M}$ due to the action of the enzyme *Cu,Zn-superoxide dismutase* (see also Chapter 5) which rapidly catalyzes the dismutation of the superoxide generated to hydrogen peroxide. This particular dismutase is located on both the stroma and lumen sides of the thylakoid membranes and the hydrogen peroxide produced is diffused to the stroma side

$$2O_2^{\cdot-} + 2H^+ \xrightarrow{\text{Cu,Zn-superoxide dismutase}} O_2 + H_2O_2$$

4.3.3 PEROXISOMES

In addition to mitochondria and chloroplasts, peroxisomes can be a source of superoxide through the activity of xanthine oxidase and NADP-requiring systems. Peroxisomes can also generate hydrogen peroxide directly through flavin oxidases as well as from the metabolism of fatty acids.

4.3.4 ENDOPLASMIC RETICULUM

In animal cells the endoplasmic reticulum (microsomes) can be another source of superoxide and hydrogen peroxide. For instance, it has been shown that the membrane-associated *cytochrome P450* complexes can yield both superoxide and hydrogen peroxide as by-products during the course of the oxidative modification of drugs or dietary toxins, as mentioned in Chapter 2. In rats exposed to the drug phenobarbitone high levels of cytochrome P450 are induced and the rate of superoxide generation from microsomes isolated from such animals can be as much as 9.6 nmol min/mg protein.

Cytochrome P450 derives its name from the characteristic absorption band around 450 nm of the ferrous iron–carbon monoxide complex. At present there are thought to be at least 60 and possibly more than 200 cytochrome P450s per mammalian species. Two discrete electron transfer reactions to cytochrome P450 are required for the turnover of the enzymes. Usually the first is NADPH via NADPH cytochrome P450 reductase which contains FAD and FMN, with a second electron from either NADPH as above or from NADH via cytochrome b_5 reductase and cytochrome b_5. All these redox sysytems are capable of super-oxide radical generation in what can again be regarded as 'accidents' of chemistry.

4.3.5 XANTHINE OXIDASE

In mammalian endothelial cells there is evidence that *xanthine oxidase* can be a natural source of superoxide radicals. Xanthine oxidase has been used by many researchers as a source of superoxide *in vitro*

$$\text{xanthine} + H_2O + O_2 \longrightarrow \text{urate} + O_2^{\cdot-}$$

Xanthine oxidase is actually derived within cells from *xanthine dehydrogenase* by proteolytic cleavage. In endothelial cells this conversion is believed to take place in response to cytokines. The dehydrogenase normally oxidizes xanthine at the expense of NADH and does not react with oxygen

$$\text{xanthine} + NAD^+ + H_2O \longrightarrow \text{urate} + NADH$$

4.3.6 ENZYMIC SOURCES OF HYDROGEN PEROXIDE

Enzymic sources of hydrogen peroxide usually involve a two-electron transfer to molecular oxygen. Examples of such enzymes are *urate oxidase, acylCoA oxidase, L-gluconolactone oxidase* and *monoamine oxidase (MAO)*. MAO catalyzes the oxidative deamination of primary aromatic amines of physiological significance (e.g. dopamine, tyramine) and is located on the outer mitochondrial membrane of mammalian cells, where it may contribute substantially to the steady-state levels of reactive oxygen species both in the mitochondrial matrix and in the cytosol.

4.3.7 SOURCES ARISING FROM DIRECT INTERACTIONS OF OXYGEN WITH CELLULAR MOLECULES

Both superoxide and hydrogen peroxide can be produced within cells as a result of the direct interaction of molecular oxygen with such important biomolecules as adrenalin, dopamine and tetrahydrofolates.

4.4 *Singlet Oxygen*

Oxygen itself is actually a radical and a good oxidizing agent. It has two unpaired electrons, each located in a different π^* antibonding orbital. These two electrons have parallel spin (the same spin quantum number). This is referred to as the ground state of oxygen and is its most stable state. More reactive forms of oxygen, called *singlet oxygens*, e.g. $^1\Delta gO_2$ and $^1\Sigma g^+O_2$, are generated by input of energy. While much more reactive than ground state oxygen, they are not radicals. They can, however, be formed in some radical reactions and can promote other radical reactions.

Singlet oxygen is most often created by *photosensitization reactions*. When certain types of molecule are illuminated by light of a particular wavelength, they absorb it and the energy raises the molecule to an excited state. The excitation energy can then be transferred to an adjacent oxygen molecule, converting it to the singlet state. Concomitantly the photosensitizer molecule returns to its ground state. Many important cellular compounds are effective sensitizers of singlet oxygen formation. These include the flavin nucleotide coenzymes, bilirubin (bile pigment), retinal, various porphyrins and chlorophylls.

The singlet oxygen produced by illumination of photosensitizers such as chlorophyll with light in the presence of oxygen can react with other molecules in the vicinity, or can attack the photosensitizer itself. Thus illuminated chlorophylls can lose their green colour as they are attacked (photobleaching). While singlet oxygen can combine chemically with other molecules, it can transfer its excitation energy to them, returning to its ground state while the other molecule enters an excited state. This latter effect is referred to as *quenching*. In many cases the two types of mechanism occur simultaneously. Biological molecules with carbon–carbon double bonds are particularly susceptible to oxidative attack by singlet oxygen, e.g. chlorophylls, carotenoids.

In terms of direct DNA damage, although the mechanisms are unclear, base modifications, such as the formation of 8-oxo-guanine, are the most frequent, whereas base loss or strand breaks are relatively minor lesions.

4.5 *Nitric Oxide*

Many cell types also generate another free radical, *nitric oxide* (NO·), which may also contribute to DNA damage through interactions with superoxide (see section 4.6). Normally nitric oxide plays a number of important biological roles. Not only is it an essential physiological mediator in a wide range of situations, but it is also a vital component of immunological defence mechanisms.

As a mediator, for example, it acts to induce the relaxation of the smooth muscle cells of the vasculature, in a process involving the specific activation of guanylate cyclase. It also inhibits the aggregation of platelets. In the central nervous system, nitric oxide is released from one or several neurones, to travel retrogradely to activate further release of glutamate from presynaptic neurones in the establishment of long-term potentiation. Evidence for an immunological function first came from experiments in which mouse macrophages could be stimulated with bacterial lipopolysaccharides to produce high levels of nitric oxide. Because of its high reactivity, nitric oxide provides an additional means of killing foreign organisms. Similar stimulatory effects could be elicited by exposure of these phagocytic cells to different cytokines. In general, however, it appears that human phagocytic cells are more

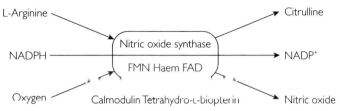

FIGURE 4.1 *The nitric oxide synthase (NOS) reaction*

LLreluctant to generate nitric oxide than are those from rodents.

The enzymes responsible for the generation of nitric oxide from the amino acid L-arginine are known as nitric oxide synthases (NOSs). They are present in almost every tissue of mammals, albeit at widely different levels, as well as in insects, snails, starfish, crabs, slime moulds and possibly plants.

The synthesis of nitric oxide requires L-arginine and NADPH and results in citrulline as well as nitric oxide. In addition the reaction requires oxygen and four cofactors (haem, FAD, FMN and tetrahydro-L-biopterin), as well as the presence of calmodulin (Figure 4.1).

Three isoforms of nitric oxide synthase can be induced and occur in many tissues and cell types.

4.6 *The Potential for Superoxide, Hydrogen Peroxide and Nitric Oxide to cause DNA Damage*

Although supcroxide, hydrogen peroxide and nitric oxide, generated either intracellularly or from adjacent inflammatory cells, have the potential to cause damage to cellular DNA, this appears to be achieved *indirectly* through interactions that propagate Fenton-type oxidants, peroxynitrite or lipid peroxides.

4.6.1 FENTON-TYPE OXIDANTS

Although superoxide itself is relatively unreactive towards DNA, it will readily dismutate, either spontaneously or in enzyme-catalyzed ractions, to yield hydrogen peroxide

$$2O_2^{\cdot-} + 2H^+ \xrightarrow{\text{dismutation}} O_2 + H_2O_2$$

While superoxide can reduce and liberate Fe^{3+} from the cellular iron-storage protein *ferritin*:

$$O_2^{\cdot-} + Fe^{3+} \longrightarrow Fe^{2+} + O_2$$

it may be more significant that superoxide can specifically liberate Fe^{2+} from cellular iron–sulphur clusters

$$[4Fe-4S]^{2+} + O_2^{\cdot-} + 2H^+ \longrightarrow [3Fe-4S]^+ + H_2O_2 + Fe^{2+}$$

The primary problem for cells is that activities that generate both *hydrogen peroxide and Fe²⁺* permit the subsequent creation of *extremely* reactive oxygen species such as *hydroxyl radicals (HO·)* through Fenton-type reactions, e.g.

$$Fe^{2+} + H^+ + H_2O_2 \longrightarrow Fe^{3+} + H_2O + HO^{\cdot}$$
$$\text{(hydroxyl radical)}$$

Much of the lethality of hydrogen peroxide towards cells is thought to involve DNA damage arising from such iron-mediated Fenton-type reactions. Within cells it is possible that NADH can drive the overall process by

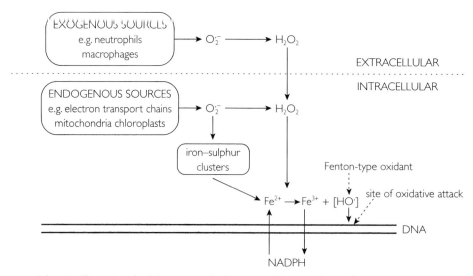

FIGURE 4.2 *Schematic illustration of cellular reactions leading to possible DNA damage by Fenton-type oxidants arising from exogenous and endogenously generated superoxide and hydrogen peroxide*

replenishing Fe^{2+} from Fe^{3+}, as well as enhancing the close association of iron with chromosomal DNA. Although it does not now seem that much of the hydrogen peroxide-induced DNA damage is actually due to diffusable hydroxyl radicals, it is believed more probable that DNA-damaging Fenton-type oxidants are created by Fe^{2+} ions intimately associated with the cellular DNA (Figure 4.2).

The precise location of the iron that binds to DNA may be critical in determining the nature of the attack. Two classes of oxidation are evident. One cleaves DNA at the following sequences: purine–*thymine*–guanine–purine (R–*T*–G–R); thymine–adenine–*thymine*–thymine–pyrimidine (T–A–*T*–Y); cytosine–thymine–*thymine*–purine (C–T–*T*–R) at the position indicated by the bases in italic. The second type makes a preferential cleavage at *purine (or pyrimidine)–guanine–guanine*–guanine (N–*G*–*G*–G). Damage by such Fenton-type oxidants can occur on either the four DNA bases themselves or the attached deoxyribose moities. The sugar damage is initiated by hydrogen abstraction from one of the deoxyribose carbons with

strand breakage and base release. Attack on the bases results primarily in OH addition to electron-rich double bonds, although hydrogen abstraction from thymine methyl groups can also occur. This attack on DNA bases generally gives rise not to strand breaks, but rather to a spectrum of as many as 50 different types of base alteration. Although very similar to the variety of base damage caused by ionizing radiation, as described in Chapter 2, there are some differences, which could result from the direct participation of iron in product formation. Among the products, the oxidized purines such as *formamidopyrimidines (Fapys)* and *8-oxo-guanine* have been well studied, as have modified pyrimidines such as thymine glycol.

4.6.2 PEROXYNITRITE

Hydrogen peroxide and superoxide also participate in metal ion-independent reactions that lead to the creation of other reactive oxygen species. For example, superoxide can react rapidly and spontaneously with nitric oxide to yield peroxynitrite anions, the protonated form of which, peroxynitrous acid, reacts efficiently with biological molecules including DNA. The combined toxicity of nitric oxide and superoxide is believed to be partly attributable to the formation of peroxynitrite. Indeed, the normal production of nitric oxide by cells may render them susceptible to superoxide damage by this means

$$NO^{\cdot} + O_2^{-} \longrightarrow ONOO^{-}$$
$$\text{peroxynitrite}$$

Reactive nitrogen intermediates such as peroxynitrite can react with a number of important cellular molecules. In the case of DNA the result is the formation of various lesions including 8-nitroguanine and 8-oxo-guanine.

4.6.3 LIPID PEROXIDATION

DNA damage from singlet oxygen may accrue by other means. For example, lipid residues with conjugated double bonds are also a particularly important target in the context of singlet oxygen attack. They are damaged with the production of lipid endoperoxides. This is an important aspect of the overall process of lipid peroxidation. Peroxidative damage to membrane lipids is potentially harmful to the cell, as it impairs the normal functioning of cellular membrane components.

Although peroxidation of membrane lipids can be brought about by singlet oxygen, a range of other reactive oxygen species can also bring about the peroxidation of membrane lipids. In fact, lipid peroxidation can be initiated by any primary free radical which has sufficient reactivity to extract a hydrogen atom from a reactive methylene group of an unsaturated fatty acid. These include hydroxyl radicals, possibly generated from Fenton-type reactions described above. Other biologically relevant radicals capable of initiation are alkoxyl

radicals (RO·), peroxyl radicals (ROO·) and alkyl radicals (R·). The formation of the initiating species is accompanied by bond rearrangement that results in stabilization of the lipid side-chain by diene conjugate formation. The lipid radical then takes up oxygen to form the peroxyl radical (ROO·). Peroxyl radicals can combine with one another, but critically they themselves are capable of abstracting hydrogen atoms from adjacent fatty acid residues in a membrane and so propagate the chain reaction of lipid peroxidation. Thus, a *single* initiation event can result in the conversion of vast numbers of fatty acid side-chains into lipid monohydroperoxides in chain reactions, as long as oxygen supplies and unoxidized fatty acid chains are available (Figure 4.3).

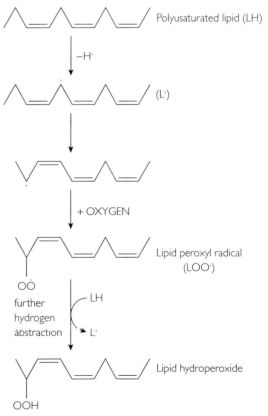

FIGURE 4.3 *Mechanism of peroxidative degradation of a polyunsaturated fatty acid*

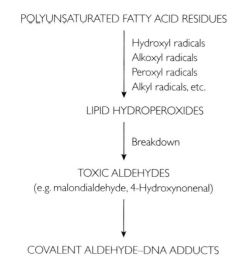

POLYUNSATURATED FATTY ACID RESIDUES

Hydroxyl radicals
Alkoxyl radicals
Peroxyl radicals
Alkyl radicals, etc.

LIPID HYDROPEROXIDES

Breakdown

TOXIC ALDEHYDES
(e.g. malondialdehyde, 4-Hydroxynonenal)

COVALENT ALDEHYDE–DNA ADDUCTS

FIGURE 4.4 *Lipid peroxidation as a possible source of DNA damage*

Lipid hydroperoxides are fairly stable molecules under physiological conditions but can decompose under certain conditions to form alkoxyl (RO$^\cdot$) and peroxyl (ROO$^\cdot$), which can reinitiate the process and amplify the initial event. Additionally, during peroxidation reactions cleavage of carbon–carbon bonds can cause the release of various toxic aldehyde products (e.g. 4-hydroxynonenal and malondialdehyde). These are hazardous to cellular components including DNA because they can form covalent aldehyde adducts (Figure 4.4).

Other environmental agents also capable of causing peroxidative damage to mammalian cells through reactions involving free radicals include certain toxic industrial chemicals (e.g. chloroform, carbon tetrachloride, ethylene dibromide, bromobenzene, halothane). For example, an initial step in the case of carbon tetrachloride (tetrachloromethane, CCl$_4$) is its metabolism by the cytochrome P450s (see Chapters 2 and 5) of the endoplasmic reticulum to yield the carbon-centred trichloromethyl radical (CCl$_3^\cdot$) which is capable of initiating the chain reactions of lipid peroxidation. The

same radical is produced from the P450 catalyzed metabolism of chloroform (trichloromethane, CHCl$_3$), although with a much reduced efficiency

$$CCl_4 \xrightarrow{\text{cytochrome P450}} {}^\cdot CCl_3 + Cl$$

4.7 *The Extent of Oxidative Damage to DNA*

A crucial question is how much oxidative damage actually occurs 'normally' to components of cellular DNA within the chromosomes of living cells. The 'steady-state' level of oxidative DNA damage from the above 'natural' sources appears surprisingly high. In humans it has been estimated at 1.5×10^5 oxidative adducts per cell. This corresponds to around 0.005% of the total number of nucleotides that comprise the human genome. Moreover, it is around one, or more, orders of magnitude higher than the steady-state level of any non-oxidative adducts in the genome. An initial approach to the determination of such 'steady-state' levels of DNA oxidation employed chemical hydrolysis of DNA isolated from cells followed by gas chromatography coupled with mass spectrometry (GC-MS). A more recent approach instead uses enzymic hydroysis of the DNA to yield nucleosides. This is followed by chromatography of the hydrolysate by HPLC. By use of electrochemical detection techniques, the adduct 8-oxo-7,8-dihydro-2-deoxyguanine has been a useful marker of oxidative DNA damage.

Difficulties can nevertheless arise from the artefactual oxidation of DNA nucleosides during DNA isolation and the analytical procedures themselves. Continual refinement of the methodology indicates that the approximate total human cellular burden of DNA oxidative adducts is as indicated above. This is based on the assumption that the 7500 8-oxo-7,8-dihydro-2-deoxyguanine adducts detected in human cell (lymphocyte) DNA represent about 5% of all oxidative adducts.

Other approaches to the detection of oxidative adducts have been pioneered. For example, monoclonal antibodies specific for specific oxidized nucleosides have been used for more precise quantification of individual adducts. A different technique is to use isolated 'DNA repair enzymes' (see Chapter 6) such as *E. coli endonuclease III* and *formidopyrimidine glycosylase (Fapy glycosylase)* which specifically recognize and excise oxidized pyrimidines. Nicks are subsequently introduced in the DNA strands at these sites, which can then be quantified and analyzed by various techniques. Overall, most data obtained by the above methods indicate that steady-state levels of chromosomal DNA damage are directly related to the basal metabolic rate of the species from which the DNA is isolated. This supports a role for the flux of damaging oxidants from mitochondria. Indeed, the level of damage to mitochondrial DNA itself is often around ten-fold higher than that of nuclear DNA. In addition, the extent of oxidative DNA damage can increase as much as two-fold with the increasing age of an animal. In short, DNA decays oxidatively with ageing.

Most recent evidence indicates that oxidative attack on chromosomal DNA components is not random and, as mentioned above, certain nucleotide sequences appear to be preferentially attacked. Moreover, attack also depends somewhat on structural arrangement of DNA and proteins within chromatin. It also appears that oxidation may be mediated by long-distance electron transport along the π stack of the DNA double helix. This suggests that the topology of DNA may serve to channel or trap oxidation in zones.

1.8 Literature Sources and Additional Reading

ASADA, K., 1992, Production and scavenging of active oxygen in chloroplasts, in Scandalios, J.G. (Ed.) *Molecular Biology of Free Radical Scavenging Systems*, pp. 173–92, Cold Spring Harbor, NY: Cold Spring Laboratory Press.

BABIOR, B.M., ElBENNA, J., CHANOCK, S.J. and SMITH, R.M., 1997, The NADPH–oxidase of leucocytes: the respiratory burst oxidase, in Scandalios, J.G. (Ed.) *Oxidative Stress and the Molecular Biology of Antioxidant Defenses*, pp. 737–83, Cold Spring Harbor, NY: Cold Spring Harbor Laboratory Press.

BECKMAN, K.D. and AMES, B.N., 1997, Oxidative decay of DNA, *Journal of Biological Chemistry*, **272**, 19633–36.

BURDON, R.H., 1995, Superoxide and hydrogen peroxide in relation to mammalian cell proliferation, *Free Radical Biology and Medicine*, **18**, 775–94.

EPE, B., 1993, DNA damage induced by photosensitisation, in Halliwell, B. and Aruoma, O.I. (Eds) *DNA and Free Radicals*, pp. 41–65, New York: Ellis Horwood.

HAHN, S.M., MITCHELL, J.B. and SHACTER, E., 1997, Tempol inhibits neutrophil and hydrogen peroxide-mediated DNA damage, *Free Radical Biology and Medicine*, **23**, 879–84.

HALLIWELL, B., 1996, Free radicals, proteins and DNA: oxidative damage versus redox regulation, *Biochemical Society Transactions*, **24**, 1023–27.

HALLIWELL, B. and GUTTERIDGE, J.M.C., 1998, *Free Radicals in Biology and Medicine*, 3rd edn, Oxford: Clarendon Press.

HAUPTMANN, N. and CADENAS, E., 1997, The oxygen paradox: biochemistry of active oxygen, in Scandalios, J.G. (Ed.) *Oxidative Stress and the Molecular Biology of Antioxidant Defenses*, pp. 1–20, Cold Spring Harbor, NY: Cold Spring Harbor Laboratory Press.

HENLE, E.S. and LINN, S., 1997, Formation, prevention and repair of DNA by iron/hydrogen peroxide, *Journal of Biological Chemistry*, **272**, 19095–98.

KNOWLES, R.G. and MONCADA, S., 1994, Nitric oxide synthases of mammals, *Biochemical Journal*, **298**, 249–58.

KWONG, L.K. and SOHAL, R.S., 1998, Substrate and site specificity of hydrogen peroxide generation in mouse mitochondria, *Archives of Biochemistry and Biophysics*, **350**, 118–26.

LAMB, C. and DIXON, R.A., 1997, The oxidative burst in plant disease resistance, *Annual Review of Plant Physiology and Molecular Biology*, 48, 251–75.

RICE-EVANS, C. and BURDON, R., 1993, Free radical–lipid interactions and their pathological consequences, *Progress in Lipid Research*, 32, 71–110.

ROQUETTE, M., PAGE, S., BRYANT, R., BENBOUBETRA, M., STEVENS, C.R., BLAKE, D.R., WISH, W.D., HARRISON, R. and TOSH, D., 1998, Xanthine oxidase is asymetrically located on the outer surface of human endothelial and epithelial cells in culture, *FEBS Letters*, 416, 397–401.

SHACTER, E., BEECHAM, E.J., COVEY, J.M., KOHN, K.W. and POTTER, M., 1988, Activated neutrophils induced prolonged DNA damage in neighbouring cells, *Carcinogenesis*, 9, 2297–304.

SOHAL, R.S., 1997, Mitochondria generate superoxide anion radicals and hydrogen peroxide, *FASEB Journal*, 11, 1269–70.

ZASTAWNY, T.H., DABROWSKA, M., JASKOLSKI, T., KLIMARCZYK, M., KULINSKI, L., KOSZELA, A., SZCZESNIEWICZ, M., SLIWINSKA, M., WITKOWSKI, P. and OLINSKI, R., 1998, Comparison of oxidative base damage in mitochondrial and nuclear DNA, *Free Radical Biology and Medicine*, 24, 722–25.

CHAPTER 5

Detoxification and Antioxidant Defences

KEY POINTS

- Organisms are equipped with a variety of enzyme systems which, by various metabolic strategies, can reduce the hazard to DNA of a range of environmental chemicals.

- The detoxification enzymes increase the water solubility of many toxic compounds as an initial step in their detoxification, metabolism and excretion. Others directly reduce the reactivity of potentially harmful compounds, and some neutralize likely toxic compounds by removing them from target cells.

- The biochemistry, genetics and molecular regulation of many of these detoxification enzymes are now well understood.

- Organisms also have an array of systems to limit the potential for free radical attack on their DNA.

- Antioxidant defence enzymes can metabolize superoxide and hydrogen peroxide and thus limit the opportunity within cells for the generation of highly reactive free radicals.

- Other defences include small molecular weight antioxidants that can readily give electrons away to free radicals and thereby reduce their reactivity towards cellular components such as DNA.

In the previous chapters, many threats to the structural integrity of cellular DNA were described. In the face of these, organisms have evolved a variety of enzymatic systems to repair the different types of DNA damage sustained. These will be discussed in Chapter 6. Organisms have also evolved an impressive range of systems to neutralize, or remove, potentially damaging compounds even before they can react with DNA. These include enzymatic systems for the detoxification of environmental chemicals as well as cellular antioxidants and antioxidant enzymes.

5.1 Systems for the Detoxification of Environmental Chemicals

Through various metabolic strategies, many *detoxification enzymes* such as cytochrome P450 oxidases, UDP–glucuronosyl transferases, N-acetyl transferases, sulphotransferases and glutathione S-transferases increase the water solubility of many toxic compounds as an initial step in their detoxification, metabolism and excretion. Others, such as the aldehyde-metabolizing enzymes, reduce the reactivity of reactive aldehydes, while ATP-dependent transporters neutralize potentially toxic compounds by removing them from target cells.

5.1.1 CYTOCHROME P450 MIXED FUNCTION OXIDASES

In humans, the metabolism of potentially toxic environmental chemicals and drugs mostly takes place in the liver. However, metabolizing capacity can also be found in other sites such as the placenta, intestinal wall, pulmonary tissues, cardiac cells and the bacteria of the gastrointestinal tract. The most important

detoxification reactions are *oxidation* and *conjugation*, although reduction and hydrolysis are also of importance in certain cases.

The enzymatic oxidation of an environmental chemical, or drug, can take place at many different molecular sites. An oxygen atom may be inserted to convert $-CH_2-$ to $-CHOH-$, or $-CH(NH_2)-$ to $-C(=O)-$. A large proportion of the oxidation reactions within cells are believed to operate under the control of the *cytochrome P450 mixed function oxidase* systems (see also Chapter 2). These systems are mainly found in the endoplasmic reticulum of eukaryotic cells, although mitochondrial P450s also exist. Hydroxylation of foreign compounds usually increases their solubility, and is a primary step in their detoxification, or metabolism and excretion.

The key overall function of the cytochrome P450 oxidases is to activate molecular oxygen for the oxidation of different types of chemical substrates. They are a large group of haem-containing proteins (with molecular weights in the range 45–55 kDa), and those in the endoplasmic reticulum require electron transfer, involving an associated NADPH–P450 reductase flavoprotein containing both flavin adenine dinucleotide (FAD) and flavin mono-nucleotide (FMN). The haem component, which acts as the catalytic centre, is attached through an iron–thiolate coordinate bond.

The main reaction catalyzed by P450s inserts a single oxygen atom into a substrate molecule. This is done by the stepwise activation and cleavage of a dioxygen molecule which also results in the formation of water. For a hydrocarbon substrate, RH, the process of mono-oxygenation via P450 can be represented as follows (the two reducing equivalents are supplied by either of the two coenzymes NADH or NADPH)

$$RH + NAD(P)H + H^+ + O_2 \longrightarrow$$
$$ROH + NAD(P)^+ + H_2O$$

An extremely intricate cycle of events is involved. The iron component of the haem starts in the ferric state, then is reduced to the ferrous state, by an electron donated from the reduced flavin mononucleotide component of the P450 reductase flavoprotein. The complex of the substrate with the ferrous form of the P450 then reacts with molecular oxygen to form a ternary complex comprising the substrate, the haem protein and oxygen, which is further reduced by another electron donated by the reduced P450 reductase flavoprotein.

During these complex reactions involving P450s there are also a number of opportunities for the generation of reactive oxygen species, such as superoxide and hydrogen peroxide, in what are essentially 'futile' cycles. This is potentially hazardous. As mentioned in Chapter 4, significant endogenous sources of these reactive oxygen species can have potentially deleterious consequences for cellular macromolecules such as DNA. Another hazardous aspect of P450 activity, as pointed out in Chapter 2, is that some of the oxidation reactions catalyzed within cells, instead of initiating the detoxification of certain environmental chemicals, can unfortunately result in their *activation*, to yield highly reactive electrophilic species capable of forming damaging adducts with cellular DNA.

As a multigene superfamily of proteins, the cytochrome P450s (CYPs) are very diverse. Current estimates of the total number of functional P450 genes in any given mammalian species range between 60 and more than 200. Based on sequence homology, the P450 superfamily in humans can be divided into ten families with each member having at least 40% homology at the amino acid level. The families CYP1, CYP2 and CYP3 are those primarily involved with the metabolism of foreign compounds such as drugs, organic solvents, anaesthetics, dyes, pesticides, alcohols, odorants, and flavours. And their diversity is believed to have arisen as an adaptive response to environmental challenge. Indeed, the level of expression of many of these genes is substrate-inducible. For example, some genes of the CYP1A subfamily are inducible by polycyclic aromatic hydrocarbons and by dioxins whereas genes in the CYP2B subfamily are regulated by polychlorinated biphenyls (PCBs) and phenobarbital. Individual P450 isoenzymes generally have distinct substrate specificities, although some overlap is observed. However, the level of their expression can vary between individuals, and an individual's response to the effects of an environmental toxin or potential carcinogen is governed by the relative activities of the P450 isoenzymes active in its metabolism. (Other P450 subfamilies located in the mitochondria rather than the endoplasmic reticulum are primarily concerned with cellular homeostasis, e.g. steroid hydroxylation reactions, oxidation of fatty acids, formation of bile acids.)

5.1.2 UDP–GLUCURONOSYL TRANSFERASES

A second type of metabolism that results in the greater water solubility of a foreign compound is conjugation. Like oxidation pathways described above, conjugation which involves coupling to another molecule requires the participation of cellular enzymes. Additionally the chemical groups involved in the coupling need to be activated through the participation of high-energy phosphate compounds. For example, glucuronic acid can be conjugated in the presence of *UDP–glucuronosyl transferase* to compounds of the general types ROH, RCOOH, RNH_2, or RSH

$$ROH + UDP–glucuronic\ acid \xrightarrow{\text{UDP–glucuronosyl transferase}} RO–glucuronide + UTP$$

Initially the glucuronic acid has to be activated. This is achieved through the reaction of glucose-1-phosphate with UTP (uridine triphosphate) catalyzed by *uridylytransferase.*

Glucose-1-phosphate + UTP ⟶
UDP–glucose + pyrophosphate

followed by the oxidation of the resulting product to activated glucuronic acid by the enzyme *UDP–glucose dehydrogenase.*

UDP–glucose + 2NAD + H_2O ⟶
UDP–glucuronic acid + 2NADH + $2H^+$

Within cells, enzyme-catalyzed conjugation can also occur with activated acetate (*N-acetyl transferases*) and sulphate (*sulphotransferases*), as well as with glutathione as described below.

5.1.3 GLUTATHIONE S-TRANSFERASES

Another important family of enzymes which catalyze conjugation are the *glutathione S-transferases* (molecular weight 17–28 kDa). They detoxify electrophilic foreign compounds, as well as cellularly generated organic halides and fatty acid peroxides, by conjugation with the cellular tripeptide glutathione (g-glutamylcysteinylglycine). Glutathione is normally abundant in cells, where it is vital for maintaining an intracellular reducing environment which ensures that intracellular protein thiol groups are not oxidized to disulphides. Additionally, it participates in the reduction of peroxides and free radicals that can accumulate in cells under oxidizing conditions.

As a component of the detoxification process, glutathione initially reacts with the compound to be detoxified (RX) in a mechanism catalyzed by glutathione S-transferase. A second enzyme, *glutamyl transpeptidase*, removes the glutamyl residue from the resulting compound. This is followed by the removal of the glycyl residue by a third enzyme, *cysteinylglycinase*, to yield an R–cysteine conjugate (Figure 5.1). This is then acetylated by an *N-acetylase* with acetylcoenzyme A to produce finally a mercapturic acid, which is more soluble and less toxic and can be secreted in the urine.

The glutathione S-transferases exist as dimers in the cytosol. There are four main classes, namely α(A), μ(M), π(P) and τ(T). Members of the same gene family show a minimum of 65% amino acid homology and can exist as either homodimers or heterodimers. In humans there are some 12 different subunits, and in rats 14. As was the case for the cytochrome P450s, the glutathione S-transferases are believed to have evolved as an adaptive response to environmental challenges and are capable of metabolizing a wide variety of structurally diverse substrates, many of which are powerful inducers of glutathione S-transferase activity. In contrast to the cytochrome P450s, glutathione S-transferases are, however, constitutively expressed at a much higher level in a wide variety of tissues, with different tissues having characteristic patterns of glutathione S-transferase isoenzyme expression.

In plants, glutathione S-transferases have been particularly studied with regard to herbicide detoxification, but they are also implicated in responses to pathogen attack, oxidative stress and heavy metal toxicity. In addition they play a role in the cellular responses to auxins and in the normal metabolism of plant secondary products such as anthocyanins and cinnamic acid.

5.1.4 ALDEHYDE-METABOLIZING ENZYMES

Aldehydes are a common environmental hazard. For instance, formaldehyde, acetaldehyde,

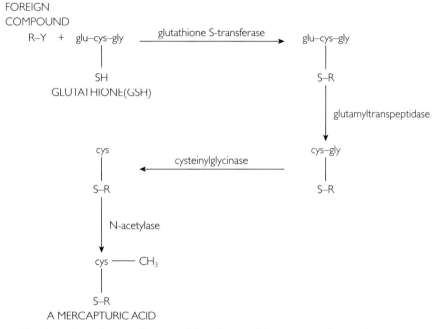

FOREIGN
COMPOUND
R–Y + glu–cys–gly   —— glutathione S-transferase ——▶   glu–cys–gly
          | |
          SH   S–R
     GLUTATHIONE(GSH)
                                                              | glutamyltranspeptidase
                                                              ▼
     cys   ◀—— cysteinylglycinase ——   cys–gly
      | |
     S–R   S–R
      |
      | N-acetylase
      ▼
     cys —— CH₃
      |
     S–R
  A MERCAPTURIC ACID

FIGURE 5.1 *Glutathione S-transferase in the process of detoxification of foreign compound RY and mercapturic acid formation (for full structure of glutathione, see Figure 5.4)*

and acrolein are products of combustion and are found in cigarette smoke and smog. It is known that formaldehyde can attack free amino groups of DNA to produce methylol derivatives. In turn these can produce cross-links between two DNA bases, or between DNA bases and amino acids of adjacent proteins. Foods, especially fruits and vegetables, are sources of aldehydes which contribute to flavours and odours. Malondialdehyde, which occurs in many foodstuffs, increases on spoilage and also in microwave-cooked red meats. A significant proportion of aldehydes are also encountered as intermediates derived from the metabolism of other compounds such as amino acids, biogenic amines, carbohydrates, vitamins, steroids and lipids. Recent interest has focused on the generation of aldehydes as the result of breakdown of peroxidised cellular lipids, e.g. 4-hydroxynonenal and malondialdehyde, which can also form DNA–aldehyde adducts (see Chapter 4). A variety of enzymes

have evolved to metabolize aldehydes to less reactive forms. These include *aldo-keto reductases*, which reduce aldehydes to alcohols, and the *aldehyde dehydrogenases*, which can oxidize aldehydes to carboxylic acids.

Aldehyde dehydrogenases (ALDHs)

The ALDHs are a family of related enzymes that catalyze the oxidation of a wide variety of aldehydes to their corresponding acids in the general reaction

$$R.CHO + NAD(P)^+ \longrightarrow$$
$$R.COOH + NAD(P)H + H^+$$

ALDHs mainly exist as tetramers of around 220–250 kDa, although some are dimeric. Some have a very wide substrate specificity, with others having very narrow substrate preferences. Three broad classes of NADP-dependent ALDHs have been observed in mammalian cells, and within each class there is evidence for multiple forms. Class 1 enzymes are cytosolic, whereas Class 2 are constitutive proteins of mitochondria. A third class is found in the cytosol of certain tumours.

Aldo-keto reductases (ALRDs)

The ALRDs are a family of enzymes that reduce a variety of aldehydes and ketones to their corresponding alcohols in the general reaction

$$R.CHO + NAD(P)H + H^+ \longrightarrow$$
$$R.CH_2OH + NAD(P)^+$$

In mammalian cells they are quite small (30–40 kDa) and are localized in the cytosol. As is the case for the ALDHs, three classes of ALRDs have been recognized, and within each class multiple forms exist.

Aldehyde reductase can reduce a range of aldehydes, especially uronic acids, some semi-aldehydes and some ketones. *Aldose reductase* overlaps somewhat in substrate specificity with aldehyde reductase, but prefers aldohexoses, converting them to corresponding polyols. *Carbonyl reductase* reduces quinones, ketones and aldehydes. Additionally, aldehyde and carbonyl reductases reduce a wide range of foreign environmental chemicals.

5.1.5 ATP-DEPENDENT TRANSPORTERS

Another means of neutralizing potentially toxic compounds is simply to remove them from target cells. In human liver there are ATP-dependent transporters capable of removing diverse damaging chemicals from cells. The *P-glycoproteins* (MDR1 and MDR2) are products of the multidrug resistance genes and are members of a family of ATP-dependent membrane-bound transporters that were originally identified in hepatocyte canicular membranes. Their main physiological function concerns the translocation of lipids and phospholipids across canicular membranes. However, the MDR1 protein can also function as an ATP-dependent efflux pump for many structurally diverse cytotoxic compounds. These include anticancer drugs, and it is their specific exclusion from cells that leads to the multidrug resistance (MDR) phenotype. With regard to possible molecular mechanisms, it is possible that MDR1, as an integral membrane protein, functions as a sort of 'hydrophobic vacuum cleaner' that can remove potentially cytotoxic chemicals or drugs directly from the plasma membrane, thus preventing their accumulation in the cytoplasm.

5.2 Cellular Antioxidant Defence Systems

Aerobic organisms gain a significant energetic advantage using molecular oxygen as the terminal oxidant in respiration. However, as pointed out in Chapter 4, oxygen has the potential to be partially reduced to reactive intermediate oxygen species. These include singlet oxygen, superoxide, hydrogen peroxide and hydroxyl radicals, which are formed not only as a result of exposure to a variety of environmental agents but also during normal cellular metabolism. Cells of aerobic organisms possess a variety of chemical and enzymatic mechanisms to protect against oxidative damage. The non-enzymatic antioxidants are generally small molecules that include, for instance, α-tocopherol, carotenoids, flavonoids and the tripeptide, glutathione, mentioned above. The enzymatic antioxidant defences comprise enzymes capable of removing, neutralizing or scavenging the reactive oxygen species; examples include superoxide dismutases, catalase, glutathione peroxidase and ascorbate peroxidase.

5.2.1 ENZYMATIC ANTIOXIDANT DEFENCES

Superoxide dismutases

Among the enzymatic mechanisms, the first line of defence is the superoxide dismutases, a family of enzymes that catalyze dismutation (a reaction in which two identical substrates have different metabolic fates). In this case one molecule of superoxide is oxidized and one is reduced

$$O_2^- + O_2^- + 2H^+ \longrightarrow H_2O_2 + O_2$$

Superoxide dismutase (SOD) is found in virtually all oxygen-consuming organisms and is able to react with superoxide at rates limited only by diffusion. In addition to possible environmental sources, considerable amounts of superoxide can arise, within both prokaryotic and eukaryotic cells, from the reduction of oxygen in electron transport chains. In eukaryotic cells, additional important sources of superoxide include the chloroplast photosystems as well as oxidases of the endoplasmic reticulum. The superoxide released from activated phagocytic lymphocytes constitutes a very significant extracellular source afflicting the invading bacteria as well as the surrounding animal cells at a site of infection. The importance of SODs is indicated by the fact that a wide range of mutant organisms lacking superoxide dismutase are hypersensitive to the toxic effects of oxygen.

Three types of superoxide dismutase have been identified. One type (Cu,Zn-SOD) contains Cu(II) plus Zn(II) at the active site, another (Mn-SOD) contains Mn(III) at the active site, while a third type (Fe-SOD) has Fe(III) at the active site. Cu,Zn-SODs (32 kDa) are found within the cytosol of eukaryotic cells as well as in the chloroplasts of plants and the periplasmic space of mainly Gram-negative bacteria. Mn-SODs are detected in mitochondria of both animals and plants. Mn-SODs (40 kDa) also occur in bacteria apparently in close association with the DNA. In contrast, Fe-SODs are generally found only in the cytosol of bacteria, although some reports suggest their occurrence in some plants and algae. The Mn-SODs and the Fe-SODs are structurally related but not to the Cu,Zn-SODs, and they may have evolved independently in response to the oxygenation of the biosphere by photosynthetic organisms. Nevertheless, the Cu,Zn-SODs and the bulk of the prokaryotic Fe-SODs and Mn-SODs exist as dimers, whereas the Mn-SODs of mitochondria are tetrameric. In plants, unlike most other organisms, there appear to be multiple forms or isoenzymes of SOD which are encoded by multiple genes.

An extracellular variety of Cu,Zn-SOD has been detected and is mainly associated with the extracellular matrix of certain mammalian cells. In humans, this enzyme, sometimes referred to as EC-SOD, is a homotetrameric protein (135 kDa) with some amino acid sequence homology to the cytosolic Cu,Zn-SOD. Its particular location positions it ideally to intercept superoxide released from phagocytic lymphocytes.

Catalases

Catalases are present in both prokaryotes and eukaryotes and catalyze the dismutation of hydrogen peroxide to water and oxygen

$$2H_2O_2 \longrightarrow 2H_2O + O_2$$

Within cells, hydrogen peroxide results from the action of the various SODs on superoxide generated from such sources as the electron transport chains of mitochondria, photosystem II components of chloroplasts or the P450 oxidases of the endoplasmic reticulum (see Chapter 4). Other intracellular sources of hydrogen peroxide include peroxisomes and reactions catalyzed by their associated oxidases and dehydrogenases. Significant extracellular sources of hydrogen peroxide are neutrophils and macrophages of inflamed tissues (see also Chapter 4). Unlike superoxide, hydrogen peroxide can readily permeate cell membranes, and a major hazard to DNA can arise from the generation of highly reactive hydroxyl radicals through the interaction of hydrogen peroxide with transition metal ions such as Fe^{2+} in Fenton-type reactions (see Chapters 2 and 4).

Structurally, most catalases (53 to 70 kDa) are made up of four protein subunits. Each individual subunit contains a haem Fe(III)–protoporphyrin component attached to its active site. In addition each subunit contains an NADPH molecule. Generally, catalases have an extremely high turnover rate (>40 000 molecules.s^{-1}) and the reaction mechanism can be represented as

$$\text{catalase–Fe(III)} + H_2O_2 \longrightarrow$$
$$\text{catalase–Fe–OOH}$$

$$\text{catalase–Fe–OOH} + H_2O_2 \longrightarrow$$
$$\text{catalase–Fe(III)} + H_2O + O_2$$

Because catalase can degrade hydrogen peroxide without consuming cellular reducing equivalents, it is a very energy efficient means of removing potentially hazardous hydrogen peroxide from cells. In the cells of plant and animal tissues, however, catalases are mainly located in peroxisomes, which also contain cellular hydrogen peroxide-generating enzymes such as gycollate oxidase, urate oxidase and the

flavin dehydrogenases (involved in the b oxidation of fatty acids). Plants, unlike animals, have multiple enzymatic forms (isoenzymes) of catalase. In maize, for example, there are three isoenzymes. CAT-1 and CAT-2 isoenzymes have been shown to be peroxisomal, whereas CAT-3 associates with mitochondria. In contrast, mitochondria from animal sources, like plant chloroplasts, have little or no catalase activity, thus the hydrogen peroxide they generate *in vivo* is unlikely to be disposed of through catalase. The situation is more puzzling in the case of microorganisms. The bacterial catalases appear to be more diverse than those from animals and plants. Moreover, it is common for bacteria to have a multiplicity of catalase isoenzymes. Although some bacterial catalases are similar to the eukaryotic enzymes, some also exhibit organic peroxidase activity, while other catalases lack haem components.

Ascorbate peroxidases

Ascorbate peroxidases are haem proteins with protoporphyrin IX as their prosthetic group. Their prime role is the rapid removal of hydrogen peroxide from sites of generation in plants and eukaryotic algae. While the hydrogen peroxide generated in peroxisomes is likely to be mainly removed by peroxisomal catalase, as mentioned above, ascorbate peroxidases can be detected in chloroplasts, as well as in the cytosol and peroxisomes.

Most leaf ascorbate peroxidase (APX) activity, however, is localized in chloroplasts and two enzymatic forms are detectable within these organelles. One is associated with the thylakoid membranes (27 kDa) and a second, slightly larger, form is found in the stroma. The cytosolic form of APX is also present in leaves, as well as in non-photosynthetic tissue. Although it is a homodimer, the molecular

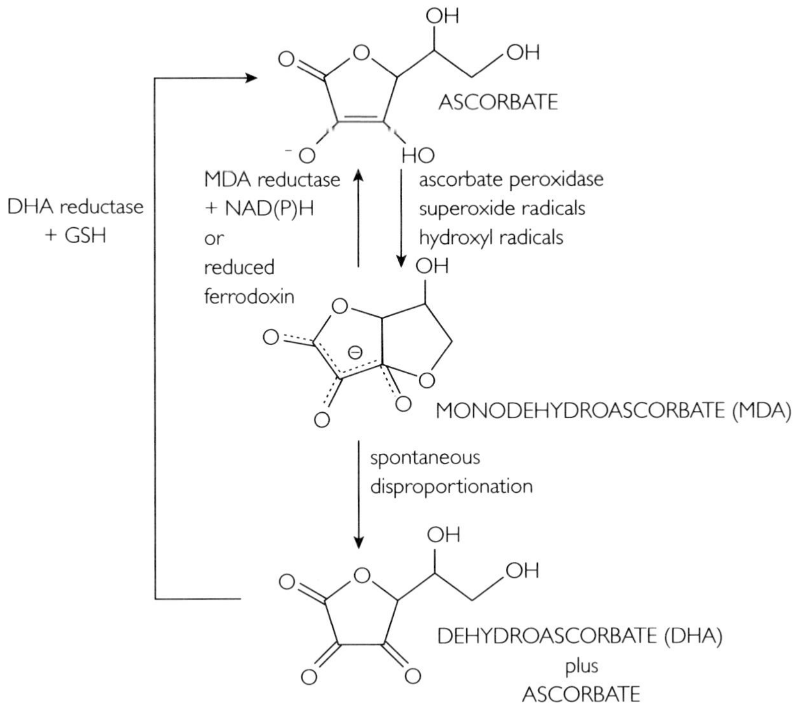

OH

OH

ASCORBATE

O   HO

DHA reductase
+ GSH

MDA reductase
+ NAD(P)H
or
reduced
ferrodoxin

ascorbate peroxidase
superoxide radicals
hydroxyl radicals

OH

O

⊖

O   O

O

O   MONODEHYDROASCORBATE (MDA)

spontaneous
disproportionation

OH

OH

O   O

O

O   O

DEHYDROASCORBATE (DHA)
plus
ASCORBATE

FIGURE 5.2 *Redox states of ascorbate*

weight of its monomer is very similar to that of the stromal APX. The peroxisomal APX is similar in size to the thylakoid APX and is mainly membrane-bound.

Peroxidases remove hydrogen peroxide through its reduction to water using electron donors (AH_2). In the case of APX, these are ascorbate and the haem group bound to the enzyme. In mammals, as will be described later in this chapter, the electron donors are glutathione and the selenocysteine residue of glutathione peroxidase.

In the reaction catalyzed by the thylakoid ascorbate peroxidase, the enzyme, initially in the ferric state, is first oxidized by hydrogen peroxide to produce a two-electron intermediate with the state of $Fe(IV)=O$ and with a tryptophan radical

$$APX(FeIII)R + H_2O_2 \longrightarrow$$
$$APX(FeIV=O)R^{\cdot} + H_2O$$

This intermediate then oxidizes ascorbate through two successive one-electron reactions, yielding two molecules of the monodehydroascorbate radical (MDA) (see Figure 5.2)

$$APX(FeIV=O)R^{\cdot} + ascorbate \longrightarrow$$
$$APX(FeIV=O)R + MDA$$

$$APX(FeIV=O)R + ascorbate \longrightarrow$$
$$APX(FeIII)R + MDA$$

Ascorbate can be regenerated from MDA with the participation of another group of plant enzymes, the *MDA reductases*. These are flavin nucleotide-containing enzymes, found in chloroplasts (55 kDa) and in the cytosol (47 kDa), as well as in mitochondria and peroxisomes. They catalyze the reduction of MDA to ascorbate by NAD(P)H

$$2MDA + NAD(P)H \longrightarrow$$
$$2 \text{ ascorbate} + NAD(P)^+$$

The monodehydroascorbate radical (MDA) can also be reduced to ascorbate by photoreduced ferredoxin in the choroplast photosytem I. Alternatively it can spontaneously dispro-portionate to ascorbate and dihydroascorbate (DHA) (see Figure 5.2), which can subse-quently be reduced by yet another enzyme, *DHA reductase*, which will also regenerate ascorbate. This enzyme of the chloroplast stroma catalyzes the reduction of DHA to ascorbate by the ubiquitous cellular tripeptide, glutathione (GSH)

$$DHA + GSH \longrightarrow ascorbate + GSSG$$

(reduced glutathione)   (oxidized glutathione or glutathione disulphide)

Glutathione peroxidases

Hydrogen peroxide within cells can also be rapidly removed from cells by glutathione peroxidases. The glutathione peroxidases (GPXs) are a family of enzymes that contain selenocysteine residues at their active centres. Selenocysteine is an unusual amino acid: it is an analogue of cysteine that contains selenium in place of sulphur, and provides a biochem-ical basis for the role of selenium in animal nutrition.

The most well-studied member of the GPX family is that found in the cytosol of ani-mal cells, which ranges in size from 76 to 100 kDa. The enzyme is homotetrameric, with each monomer of 198 amino acid residues having a selenocysteine residue at position 45. It has a broad specificity for hydroperoxide sub-strates, but cannot act on the hydroperoxides of complex lipids. However, a phospholipid hydroperoxide GPX does exist that, in addition to acting on low-molecular-weight hydroper-oxides, can also act on the hydroperoxides of lipids and cholesterol. Extracellular tetrameric GPXs have also been detected in mammalian plasma. Although GPX-like activities have been reported recently in certain algae and plants, these remain to be fully characterized.

With regard to the catalytic mechanism of GPX, although it is still only partially under-stood, the specificity towards glutathione as substrate is absolute. Hydrogen peroxide or lipid hydroperoxides (LOOH) are removed through their reduction to water, using the electron donors glutathione (GSH) and the selenocysteine residue of the enzyme

$$H_2O_2 + GSH \xrightarrow{GPX} 2H_2O + GSSG$$

$$LOOH + GSH \xrightarrow{GPX} LOH + H_2O + GSSG$$

Glutathione reductases

Usually both glutathione peroxidases and ascorbate peroxidases operate in cycles with another group of enzymes, the *glutathione reduc-tases*, which use reducing equivalents, derived from glucose through the pentose phosphate pathway and NADPH, to regenerate the reduced form of glutathione (GSH) from the oxidized disulphide form (GSSG) that results from the action of glutathione peroxidases or ascorbate peroxidases as already described

$$GSSG + NADPH + H^+ \xrightarrow{\text{glutathione reductase}} 2GSH + NADP$$

Glutathione reductases are found in bacteria and in the cytosol and mitochondria of plant and animal cells, as well as in plant chloro-plasts. They range in size from around 90 to 140 kDa and usually contain two protein sub-units, each with a flavin dinucleotide (FAD) at its active site. It appears that the NADPH reduces the flavin nucleotide, which then transfers its electrons onto a disulphide bridge (–S–S–) in the enzyme. The two sulphydryl groups (–SH) that result then interact with glutathione disulphide (GSSG) and reduce it to GSH. The activity of glutathione reductase thus ensures that the ratio of GSH/GSSG in normal cells is kept high.

5.2.2 NON-ENZYMATIC ANTIOXIDANT DEFENCES

α-Tocopherol (vitamin E)

One of the major cellular targets for attack by oxygen-derived free radical species is the polyunsaturated fatty acid components of phospholipid cell membranes. This can lead to the propagation, in chain reactions, of extensive peroxidative damage (see Chapter 4). However, the presence in the membranes of a lipid-soluble chain-breaking antioxidant such as α-tocopherol (vitamin E) can quench this propagation. Oxidation of α-tocopherol to α-tocopheroxy radicals occurs during this process when lipid peroxyl radicals are quenched

(Figure 5.3). These products, however, can accept hydrogen to regenerate the original α-tocopherol molecules (Figure 5.3). Recent interest has focused on the possibility that ascorbate (vitamin C) might be involved in this regeneration mechanism.

Primary sources of vitamin E are plant tissues. Particularly rich, from a human dietary point of view, are vegetable oils, nuts and whole grains. Vitamin E deficiency disease is common among farm animals and can cause considerable reproductive problems. In humans, however, acute vitamin E deficiency is rare as most diets contain at least some vitamin E.

Ascorbic acid (vitamin C)

Besides functioning as an electron donor in the enzymatic removal of hydrogen peroxide catalyzed by ascorbate peroxidase, as described in the previous section, water-soluble ascorbate can be oxidized non-enzymatically by superoxide or hydroxyl radicals to yield the monodehydroascorbate radical (MDA) (see Figure 5.2).

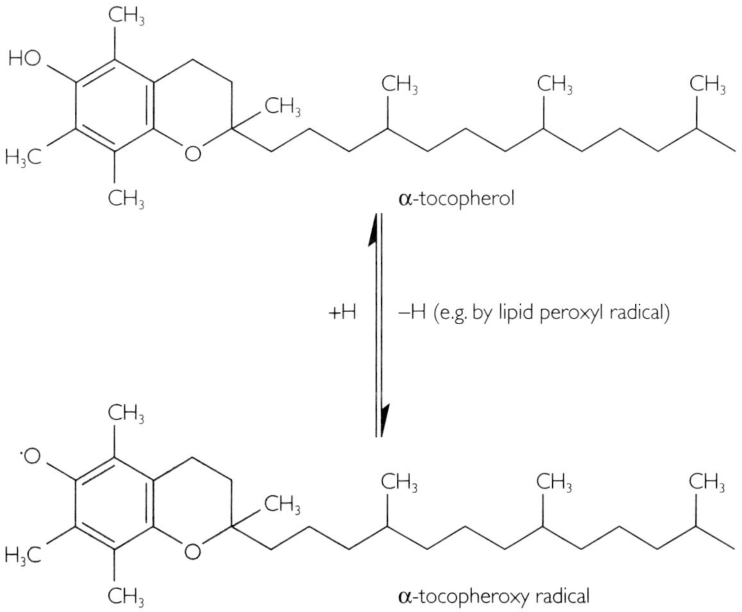

FIGURE 5.3 *Oxidation and regeneration of α-tocopherol*

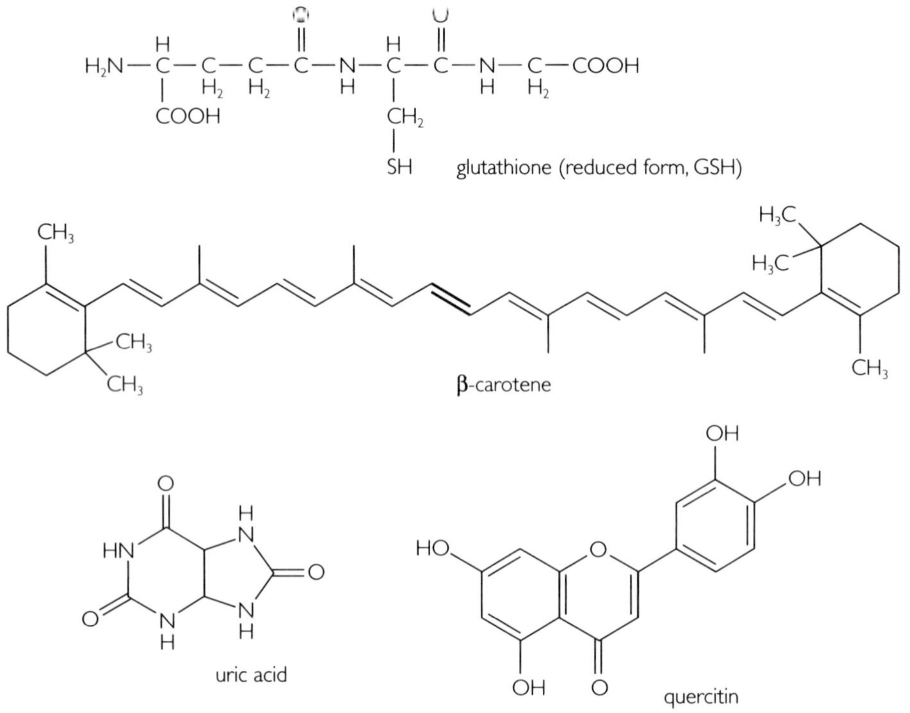

FIGURE 5.4 *The structures of glutathione, β-carotene, uric acid and quercitin*

In this sense ascorbate can act directly as a scavenger of superoxide or hydroxyl radicals. Ascorbate can be regenerated from the resulting MDA through the activity of MDA reductase (Figure 5.2).

Ascorbate, or vitamin C, is abundant in plant tissues. From a dietary point of view, particularly rich sources are blackcurrants, peas, potatoes, Brussels sprouts, broccoli, oranges and kiwi fruit. Vitamin C is required in the diet of primates (including humans) and guinea pigs. This is due to the lack of a gene that encodes one of the enzymes required for ascorbate synthesis from glucose.

Glutathione

As detailed in the preceding section, the widely distributed cellular tripeptide glutathione (GSH) (Figure 5.4) has an important role in protecting cells from potential damage from

reactive oxygen species by providing reducing equivalents for several of the key antioxidant defence enzymes such as ascorbate peroxidase and glutathione peroxidase. Glutathione can also scavenge hydroxyl radicals directly. This process, however, results in the formation of glutathiyl radicals (GS·), which themselves may initiate radical reactions, although they are less reactive than hydroxyl radicals

$$GSH + HO· \longrightarrow GS· + H_2O$$

Thus, although the predominant function of glutathione in living cells is protective, the possibility of undesirable secondary reactions cannot be overlooked.

Uric acid

Uric acid (Figure 5.4) is another small water-soluble molecule that may have significance in the provision of antioxidant defence. It accumulates in human tissues as the end-product of purine metabolism and is a potent scavenger of peroxyl and hydroxyl radicals as well as singlet oxygen.

Carotenoids

In plants, carotenoids are important scavengers of singlet oxygen which is generated by an excited state of chlorophyll, whose existence is also decreased by carotenoids. This protective mechanism takes place by energy transfer from the excited chlorophyll and singlet oxygen to the carotenoid, which absorbs and dissipates it without chemical change. In animals the role of carotenoids is unclear, although singlet oxygen is produced during the irradiation of skin cells with ultraviolet light. The structure of β-carotene is shown in Figure 5.4. More than 600 naturally occurring carotenoids have now been identified.

Flavonoids and polyphenolics

In addition to ascorbate, tocopherols and carotenoids there are many other plant compounds that have the ability to scavenge free radicals. For instance, there are many polyphenolic flavonoids with a basic diphenylane-propane skeleton; the family includes flavanols, flavanones, flavones, flavonols, and anthocyanadins. An example of a flavonol, quercitin, is shown in Figure 5.4. Quercitin is a strong antioxidant which can prevent the free radical-induced peroxidation of lipids *in vitro* and is believed to protect against heart disease *in vivo*. Good sources are apples, onions and tea. Green tea polyphenols can inhibit the oxidant-induced DNA strand breakage in cultured human lung cells.

5.3 Literature Sources and Additional Reading

ASADA, K., 1997, The role of ascorbate peroxidase and monodehydroascorbate reductase in H_2O_2 scavenging in plants, in Scandalios, J.G. (Ed.) *Oxidative Stress and the Molecular Biology of Antioxidant Defenses*, pp. 715–35, Cold Spring Harbor, NY: Cold Spring Harbor Laboratory Press.

CHAPPLE, C., 1998, Molecular–genetic analysis of plant cytochrome P450-dependent monooxygenases, *Annual Review of Plant Physiology and Plant Molecular Biology*, 49, 311–43.

CHIN, K.-V., PASTAN, I. and GOTTESMAN, M.M., 1993, Function and regulation of the human multidrug resistance gene, *Advances in Cancer Research*, 60, 157–80.

COON, M.J., DING, X., PERNECKY, S.J. and VAZ, A.D.N., 1992, Cytochrome P450: progress and predictions, *FASEB Journal*, 6, 669–73.

DANIEL, V., 1993, Glutathione S-transferases: gene structure and regulation of expression, *Critical Reviews in Biochemistry and Molecular Biology*, 28, 173–207.

DIPLOCK, A.T., 1994, Antioxidants and free radical scavengers, in Rice-Evans, C.A. and Burdon, R.H. (Eds) *Free Radical Damage and its Control*, pp. 113–30, Amsterdam: Elsevier Science.

ESTABROOK, R.W., WEISS, R.H., WERRINGLOER, J. and PETERSON, J.A., 1991, The cytochrome P450s and oxygen stress, in Davies, K.J.A. (Ed.) *Oxidative Damage and Repair: Chemical, Biological and Medical Aspects*, pp. 720–25, Oxford: Pergamon Press.

FLOHE, L., WINGENDER, E. and BRIGELIUS-FLOHE, R., 1997, Regulation of glutathione peroxidases, in *Oxidative Stress and Signal Transduction*, pp. 415–40, New York: Chapman and Hall.

FRIDOVICH, I., 1997, Superoxide anion radical (O_2^-), superoxide dismutases and related matters, *Journal of Biological Chemistry*, 272, 18515–17.

GLATT, H., 1997, Bioactivation of mutagens via sulphation, *FASEB Journal*, 11, 314–21.

GONZALEZ, F.J. and NEBERT, D.W., 1990, Evolution of the P450 superfamily: animal–plant warfare, molecular drive and human genetic differences in drug oxidation, *Trends in Genetics*, 6, 182–86.

GRALLA, E.B., 1997, Superoxide; studies in the yeast *Saccharomyces cerevisiae*, in Scandalios, J.G. (Ed.) *Oxidative Stress and the Molecular Biology of Antioxidant Defenses*, pp. 495–525, Cold Spring Harbor, NY: Cold Spring Harbor Laboratory Press.

HALLIWELL, B. and GUTTERIDGE, J.M.C., 1998, *Free Radicals in Biology and Medicine*, 3rd edn, Oxford: Clarendon Press.

LEANDERSON, P., FARESJO, A.O. and TAGESSON, C., 1997, Green tea polyphenols inhibit oxidant-induced DNA strand breakage in cultured lung cells, *Free Radical Biology and Medicine*, 23, 235–42.

LEWIS, D., 1997, Sex, drugs and P450, *Chemistry and Industry*, 20 October, 831–34.

LINDAHL, R., 1992, Aldehyde dehydrogenases and their role in carcinogenesis, *Critical Reviews in Biochemistry and Molecular Biology*, 27, 283–335.

LOEWEN, P.C., 1997, Bacterial catalases, in Scandalios, J.G. (Ed.) *Oxidative Stress and the Molecular Biology of Antioxidant Defenses*, pp. 273–308, Cold Spring Harbor, NY: Cold Spring Harbor Laboratory Press.

MARRS, K.A., 1996, The functions and regulation of glutathione S-transferases in plants, *Annual Review of Plant Physiology and Molecular Biology*, 47, 127–58.

MOREL, F., SHULTZ, W.A. and SIES, H., 1994, Gene structure and regulation of expression of human glutathione S-transferase alpha, *Biological Chemistry Hoppe-Seyler*, 375, 641–49.

MULLINEAUX, P.M. and CREISSEN, G.P., 1997, Glutathione reductase: regulation and role in oxidative stress, in Scandalios, J.G. (Ed.) *Oxidative Stress and the Molecular Biology of Antioxidant Defenses*, pp. 667–713, Cold Spring Harbor, NY: Cold Spring Harbor Laboratory Press.

NOCTOR, G. and FOYER, C.H., 1998, Ascorbate and glutathione: keeping active oxygen under control, *Annual Review of Plant Physiology and Plant Molecular Biology*, 49, 249–79.

RICE-EVANS, C.A., MILLER, N.J. and PAGANGA, G., 1996, Structure–antioxidant activity relationships of flavonoids and phenolic acids, *Free Radical Biology and Medicine*, 20, 933–56.

RUIS, H. and KOLLER, F., 1997, Biochemistry, molecular biology and cell biology of yeast and fungal catalases, in Scandalios, J.G. (Ed.) *Oxidative Stress and the Molecular Biology of Antioxidant Defenses*, pp. 309–342, Cold Spring Harbor, NY: Cold Spring Harbor Laboratory Press.

SCANDALIOS, J.G., 1997, Molecular genetics of superoxide dismutases in plants, in Scandalios, J.G. (Ed.) *Oxidative Stress and the Molecular Biology of Antioxidant Defenses*, pp. 527–68, Cold Spring Harbor, NY: Cold Spring Harbor Laboratory Press.

SCANDALIOS, J.G., GUAN, L. and POLIDOROS, A.N., 1997, Catalases in plants: gene structure, properties, regulation and expression, in Scandalios, J.G. (Ed.) *Oxidative Stress and the Molecular Biology of Antioxidant Defenses*, pp. 343–406, Cold Spring Harbor, NY: Cold Spring Harbor Laboratory Press.

SILVERMAN, J.A. and SCHRENK, D., 1997, Expression of the multidrug resistance gene in the liver, *FASEB Journal*, 11, 308–13.

SMITH, G., SMITH, C.A.D. and WOLF, C.R., 1995, Pharmacogenetic polymorphisms, in Phillips, D.H. and Venitt, S. (Eds) *Environmental Mutagenesis*, pp. 83–106, Oxford: Bios Scientific.

TUATI, D., 1997, Superoxide dismutases in bacteria and pathogen protists, in Scandalios, J.G. (Ed.) *Oxidative Stress and the Molecular Biology of Antioxidant Defenses*, pp. 447–93, Cold Spring Harbor, NY: Cold Spring Harbor Laboratory Press.

WINGARD, L.B., BRODY, T.M., LARNER, J. and SCHWARTZ, A., 1991, *Human Pharmacology, Molecular to Clinical*, pp. 50–63, London: Wolfe Publishing.

CHAPTER 6

DNA Repair Mechanisms

KEY POINTS

■ Besides the damaging effects to DNA of a wide range of environmental factors, there is an inherent potential for error during the DNA replication process itself, as well as the occurrence of spontaneous lesions.

■ In the face of these problems living organisms have evolved an impressive series of enzyme systems that repair DNA damage in various ways.

■ Some enzymes, such as the photolyases and alkyl transferases, can reverse DNA damage directly.

■ More complex excision repair systems are available to remove damage to bases.

■ In the case of DNA replication errors, post-replication mismatch repair systems exist to eliminate gaps that arise in DNA strands at positions directly opposite lesions that block DNA replication.

■ Mechanisms are also present that facilitate the repair of double-strand DNA breaks.

The spectrum of damage that cellular DNA can suffer is extensive. Although various types of environmental factor are hazardous because they can bring about specific damage to DNA, considerable DNA damage appears to arise from the activities of reactants generated naturally within cells themselves. In the face of this range of problems, organisms have evolved an impressive array of enzymatic repair systems to restore the normal nucleotide sequence and structure of damaged DNA. Moreover, these systems are capable of repairing damage caused by both synthetic and naturally produced reactants. This chapter aims to review very broadly the various types of enzyme-based repair mechanisms that exist. These include the photolyases and alkyl transferases, which are enzymes that can reverse DNA damage directly, as well as the more complex nucleotide and base excision systems that require the participation of considerable numbers of enzymes. In addition, post-replication mismatch repair systems that can eliminate gaps arising in DNA strands at positions directly opposite damaged nucleotides that block DNA replication are described, together with mechanisms available to facilitate the repair of double-strand DNA breaks.

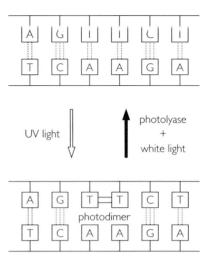

FIGURE 6.1 *Repair of an ultraviolet-induced pyrimidine photodimer in DNA by a photolyase*

pyrimidine bases in the DNA (Figure 6.1). The enzyme cannot function in the dark and is not present in human cells. However, in plants there are photolyases capable of cleaving 6−4 photoproducts as well as cyclobutane dimers (see Figure 2.11). Although photoreactivation is undoubtedly the plant kingdom's major line of defence against UV-induced damage, the biochemical nature of the two types of photolyase involved is unknown.

6.1 Photolyases

The most straightforward means of repair is the direct reversal of DNA damage. Although most types of DNA damage cannot be directly reversed, there are photolyases in bacteria and lower eukaryotes, which can directly repair cyclobutane pyrimidine dimers formed by ultraviolet radiation (see Figure 2.11). The photolyase enzyme binds to the photodimer and simply splits it, in the presence of certain wavelengths of light, to generate the original

6.2 Alkyltransferases

Another set of enzymes capable of directly reversing DNA damage is the alkyltransferases. For example, they can remove certain alkyl groups from the O−6 position of guanine that have been added by alkylating agents such as ethane methane sulphonate (see Figure 2.3). In E. coli, the enzyme transfers the methyl group from the O−6-methyl guanine residue in DNA to one of its own cysteine residues (Figure 6.2). Because this step inactivates the alkyltransferase molecule, the

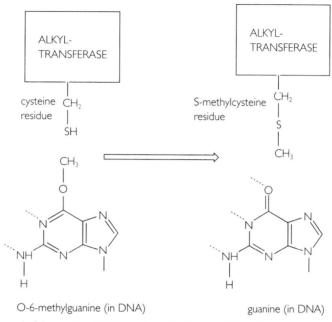

cysteine residue

S-methylcysteine residue

O-6-methylguanine (in DNA)

guanine (in DNA)

FIGURE 6.2 *Direct reversal of damage to a DNA guanine residue by an alkyltransferase*

amount of alkylation damage that this type of repair system can cope with is clearly limited by the level of availability of unreacted alkyltransferase molecules.

6.3 *Nucleotide Excision Repair*

In nucleotide excision repair (NER), damaged DNA is removed as a part of an oligonucleotide fragment, followed by replacement with new DNA using the remaining intact strand as a template. It is a complex process involving the participation of possibly as many as 30 different proteins, but with a broad specificity. A wide range of chemical alterations to DNA that result in significant local distortions of DNA structure are recognized by this now well-studied repair system. Such alterations include some types of oxidative damage as well as the different types of photodimer induced

by ultraviolet radiation. Figure 6.3 illustrates the basic action of the NER system towards a photodimer that might arise in bacterial DNA after ultraviolet exposure.

The process involves the breaking of a phosphodiester bond on either side of the damaged site on the same strand, resulting in the excision of an oligonucleotide. This leaves a gap in one of the DNA strands which is filled in by *DNA repair synthesis*. A DNA *ligase* then seals the remaining break (Figure 6.3). In prokaryotes the size of the fragment removed is 12 or 13 nucleotides, whereas in eukaryotes it is usually 27 to 29 nucleotides in length.

In *E. coli*, the excision of the oligonucleotide fragment requires the coordinated activity of three proteins. One of these, the UvrA protein, which can recognize the damaged portion of DNA, combines with another protein, UvrB, to take it to the site of damage. Having accomplished that task, the UvrA protein then severs its association with UvrB

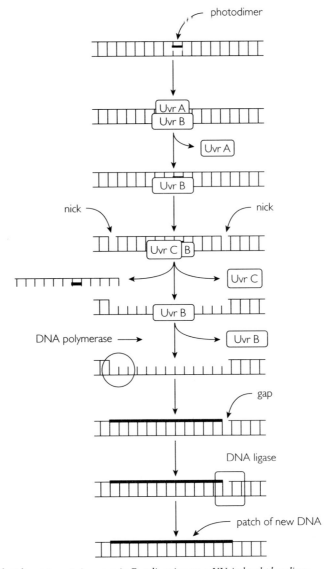

FIGURE 6.3 *The nucleotide excision repair system in E. coli acting on a UV-induced photodimer*

and another protein, UvrC, complexes with UvrB at the site of DNA damage. This complex then acts as an *exonuclease* to catalyze the nicking of the DNA strand on either side of the damage. With the help of a *helicase* protein, this complex facilitates the physical removal of the oligonucleotide fragment containing the damaged nucleotides (Figure 6.3). To fill in the gap left by this process, the enzyme *DNA polymerase I* synthesizes new DNA in the direction 5′ to 3′, using the intact strand that remains as its template (Figure 6.3). After the final gap is sealed by *DNA ligase*, the patching up of the DNA is complete.

Although the bacterial exonuclease comprises only three proteins, its human counterpart is more complicated, with at least

17 different proteins. The overall process in human cells is nevertheless very similar to that in bacteria and the repair synthesis phases appear to involve DNA polymerase δ and ε, together with a protein called the proliferating cell nuclear antigen (PCNA). A unique NER pathway for repair of 8-oxo-guanine lesions has been described within human cells which makes incisions immediately to either side, leaving only a one nucleotide gap, to be made good by subsequent DNA repair synthesis systems.

In both humans and bacteria, it has emerged that there is preferential repair of the particular DNA strands that are transcribed when genes are active (see Chapter 7 for gene transcription mechanisms). The coupling between the NER system and transcription is mediated through a partnership with a basic transcription factor designated TFIIH (see Chapter 7). This transcriptional regulatory protein, which participates in the unwinding of DNA, allows certain of the repair nucleases to gain access to the sites of damage within active genes. Besides the connection with gene transcription, a major impetus to the study of NER systems is the finding that deficiencies in the human NER system have been linked to the highly cancer-prone genetic disorder *xeroderma pigmentosum*, as well as to other disorders such as *tricothiodystrophy* and *Cockayne syndrome*.

6.4 *Base Excision Repair Pathways*

Base excision repair (BER) pathways are likely to have evolved to protect cells from the damaging effects to DNA bases brought about by spontaneous hydrolysis, or by endogenously produced reactive oxygen species, or other cellular metabolites that can modify DNA. BER, however, is also important for the rectification of similar base damage induced by ionizing radiation and alkylating agents. Key enzymes in this type of pathway are the *DNA glycosylases*. Different DNA glycosylases are required to remove distinct varieties of damaged base, and the specificity is dictated by the species of glycosylase involved. A variety of different DNA glycosylases have now been identified in human and bacterial cells, some examples of which are listed in Table 6.1.

DNA glycosylases cleave the bonds between bases and deoxyribose residues (N-glycosylic bonds), releasing the damaged bases and generating an *apurinic*, or *apyrimidinic site*, referred to collectively as *AP sites* (Figure 6.4). Another related group, the *glycosylase/AP lyase* enzymes, cleave the N-glycosylic bond as well as the DNA–phosphate backbone, to leave a nick with an unsaturated sugar at the 3′-terminus and a 5′-phosphate group, but their function is unclear.

TABLE 6.1 *Some DNA Glycosylases*

Thymine glycol DNA glycosylase
8-oxo-guanine/ Fapy* DNA glycosylase
3-methyl adenine/3-methyl guanine/ 7-methyl guanine/ hypoxanthine DNA glycosylase
Hydroxymethyl uracil DNA glycosylase
Formyl uracil DNA glycosylase
Uracil glycosylase

* Fapy, formamidopyrimidine.

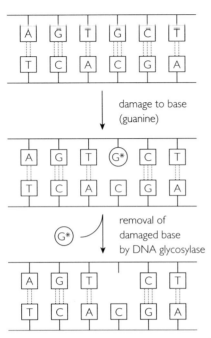

FIGURE 6.4 *Removal of a damaged DNA guanine by a DNA glycosylase to yield an apurinic site*

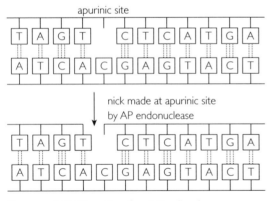

FIGURE 6.5 *The action of an AP endonuclease*

The AP sites that result from DNA glycosylase action are then attacked by *AP endonucleases* that exist in all cells, with the resultant cleavage of the phosphodiester bonds at the AP site (Figure 6.5). These enzymes initiate sugar removal by hydrolyzing the 5′-phosphodiester bond of an AP site. The 5′-terminal deoxyribose that results is then

a substrate for a *deoxyribosephosphodiesterase* activity that appears to be associated with DNA glycosylases. Once the baseless deoxyribose is removed, the small gap that results is filled, most likely by DNA polymerase I in bacteria, or by DNA polymerase β in higher eukaryotes, and then sealed by DNA ligase. Another role for the deoxyribosephosphodiesterase activity may well be in the removal of deoxyribose fragments, which can arise in DNA following exposure to ionizing radiation (see Figure 2.9).

Although AP sites result primarily from the removal of damaged bases following the action of DNA glycosylases, it will be recalled from Chapter 3 that depurination is quite a frequent spontaneous event. This would also yield an AP site, but one which can be repaired in the manner described above. Deamination of DNA bases is also a frequent spontaneous event, but the products, hypoxanthine, xanthine and uracil, can nevertheless be removed through base excision pathways involving specific DNA glycosylases.

6.5 *Mismatch Repair*

DNA mismatch repair systems exist to repair mispaired bases formed during DNA replication, genetic recombination or as a result of damage to DNA.

In *E. coli* the major mismatch repair system is the *MutHLS* system. The nucleotide sequence GATC in the DNA of *E. coli* is normally 'methylated', i.e. the adenine residues are methylated enzymically at the N-6 position through the action of an *adenine methylase* (Figure 6.6). However, immediately following replication the newly synthesized GATC sequence in the daughter strand is briefly in a hemi-methylated state (Figure 6.6). This particular lack of base methylation serves as

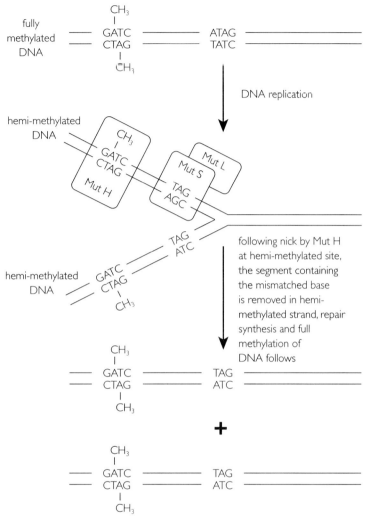

FIGURE 6.6 *A schematic illustration of the action of the MutHLS mismatch repair system in E. coli*

a type of specific 'signal' which distinguishes the newly synthesized DNA strand from the parental strand. In the example presented in Figure 6.6, an error has occurred. A guanine nucleotide has been inserted in the new strand rather than the correct thymine nucleotide.

Basically, the repair activity involves the mismatch-dependent nicking of the unmethylated strand at a hemi-methylated GATC site and degradation from the nick past the mismatch, followed by resynthesis. The MutS protein binds to DNA at the site of of the mispaired base (see Figure 6.6). Mut L protein then interacts with with MutS and is necessary for the activation of a further protein, MutH. MutH is an endonuclease that nicks hemi-methylated DNA in the unmethylated strand and is activated by MutS and MutL in the presence of a mismatch (Figure 6.6). The subsequent excision reactions require the participation of exonucleases and a helicase. Once excision is complete, resynthesis of the missing DNA is achieved by DNA polymerase III and DNA ligase.

Additionally in *E. coli* there exists the *MutY* system. If a sequence of DNA containing the oxidized base 8-oxo guanine is not repaired by the base excision pathway involving the appropriate glycosylase (see Table 6.1), it may be replicated and an adenine (A) nucleotide is sometimes misincorporated opposite the 8-oxo-guanine which results in the change of a G–C base pair for a T–A at the next round of replication. Mispairing between adenine and 8-oxo-guanine is rectified by a reaction in which MutY acts not only as a glycosylase which removes the adenine, but also as an AP nuclease that makes a nick adjacent to the AP site. Excision and resynthesis is then achieved with DNA polymerase I and DNA ligase.

A mismatch repair system with some similarities has been detected in human cells. For example, two components of this system hMSH2 and hMLH1 are very similar to the bacterial proteins MutS and MutL respectively. Although the nature of the 'strand signal' has not yet been identified in any eukaryotic organism, mismatched bases appear to be recognized by a complex of hMSH2 with a second MutS homologue. This is followed by the binding of a complex containing hMLH1, which then recruits other proteins involved in the subsequent steps of excision and DNA resynthesis. Such proteins include DNA polymerase δ and PCNA (the proliferating cell nuclear antigen). Mutated versions of hMSH2 and hMLH1 are prevalent in *hereditary non-polyposis colon cancer (HNPCC)*, which is one of the most common human cancer susceptibility syndromes known. Although colon cancers are the most prevalent types of tumour seen in HNPCC families, approximately 35–40% of HNPCC families develop other types of cancer and it now appears the HNPCC could be associated with inherited mismatch repair system defects.

6.6 *Error-prone Replication*

In bacteria there exists an '*SOS system*', so called because it appears to be induced as an emergency response to forestall cell death in the event of significant damage to DNA. It is very much a last resort and is essentially a trade-off between survival and the subsequent presence of mismatched nucleotides in its DNA. The replication system is allowed to continue past the site of DNA damage and in the process is able to accept as substrates any nucleotides, complementary or not, in an effort to achieve synthesis of complete new strands of DNA (Figure 6.7). Precisely how this 'bypass' mechanism functions is not clear, but in *E. coli* at least three proteins including RecA, UmuC and UmuD appear to be involved. The last two probably combine with the DNA polymerase replication complex to lessen its strict specificity for only the nucleotide substrates that are complementary, and thus permit replication past the sites of damage.

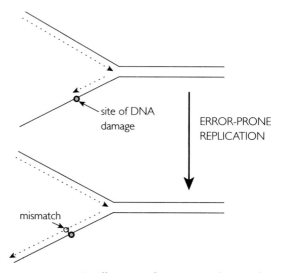

FIGURE 6.7 *An illustration of error-prone replication of DNA characteristic of the SOS response*

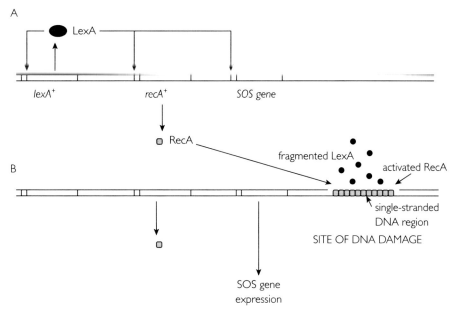

FIGURE 6.8 *A diagrammatic representation of the roles of LexA and RecA proteins in regulating the expression of SOS response in E. coli (A, the uninduced situation; B, after DNA damage)*

Normally in *E. coli* the protein product of the *lexA*$^+$ gene acts as a repressor of more than 20 different cellular genes, which include *recA*$^+$ as well as *lexA*$^+$ itself, by binding to similar regulatory sequences associated with each gene (Figure 6.8). By this means a low level of expression of *lexA*$^+$ and *recA*$^+$ genes is maintained. Indeed, it has been estimated that around 7000 RecA protein molecules are normally present in each bacterium and that these play a key role in general genetic recombination processes. Damage to DNA, however, signals the induction of SOS responses. The actual signal is believed to be regions of single-stranded DNA that are generated when cells attempt to replicate a damaged DNA template, or when the replication process is impeded. The RecA protein then binds to these regions, and in the process is converted to an activated form. LexA protein molecules then associate with this activated RecA and as a result are cleaved in a process which subsequently facilitates their autodigestion. By this means the cellular levels of LexA protein are reduced, which then allows the various SOS response genes, including *recA*$^+$, to be expressed at enhanced levels.

6.7 *Recombination Repair*

Recombination is a fundamental process in biology whereby genetic material from one parental chromosome and from the other parental chromosome are effectively 'cut and pasted', for example at meiosis. In eukaryotes, recombination can occur at regions of nucleotide sequence homology between the two chromosomes, through the physical breakage and rejoining of DNA molecules. It is now clear that biochemical reactions associated with the basic processes of genetic recombination are also involved in the repair of both double-strand and single-strand gaps in damaged DNA. Additionally, complex lesions such as interstrand cross-links may also be removed by *recombination repair pathways*.

6.7.1 REPAIR OF DOUBLE-STRAND BREAKS

Work with yeasts and bacteria has suggested that, under certain conditions, two homologous DNA molecules are necessary for the repair of damaged DNA, and various models for the repair of double-strand breaks have been proposed. The first step in one such model for the post-replication recombination repair of a double-strand break is the enzymic processing of the DNA ends that results from the double-strand break, to yield 3′-OH single-stranded 'tails' (Figure 6.9). This is followed by a 'synaptic' phase which involves a search

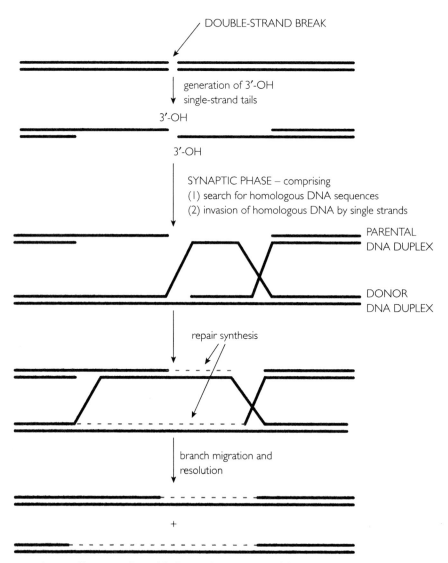

FIGURE 6.9 *A schematic illustration of a model of post-replication repair of double-strand break*

for homologous DNA sequences in a sister DNA duplex, and the ensuing invasion of the homologous double-stranded donor DNA by single-stranded tails generated from the damaged double-stranded DNA. Repair synthesis is then required to fill gaps in both DNA molecules, followed by branch migration and complex junction resolution steps (Figure 6.9).

At least 20 gene products have been shown to be involved in the complex biochemistry of normal genetic recombination in *E. coli*, and many of these also play key roles in recombination repair processes. One of these, the RecBCD protein, has both exonuclease and DNA helicase activities and catalyzes the formation of the 3′-OH single-stranded tails at the double-strand breaks in damaged DNA molecules. To achieve this, it unwinds the double-stranded DNA ends at the break, and then degrades them from their 5′-ends to leave 3′-OH terminated single-stranded tails. Another recombination protein, RecA, is responsible for carrying out a homology search followed by stand exchange reactions. In addition to its role in SOS responses, RecA is able to pair two homologous DNA molecules if one is single-stranded, or partially single-stranded. The active species in this process is a right-handed helical nucleoprotein filament composed of DNA and units of the RecA protein that can be thought of as a scaffold to facilitate DNA pairing and strand exchange. RecA can form filaments with both single- and double-stranded DNA.

6.7.2 REPAIR OF SINGLE-STRAND GAPS

Homologous recombination can also assist in the repair of single-stranded gaps that form opposite damaged DNA regions. For example, the normal DNA replication machinery will 'stall' at bulky base adducts, or photodimers

induced by ultraviolet radiation, and then 'restart' the replication process past such blocks, thus leaving a single-stranded gap (Figure 6.10). In *post-replication recombination repair*, this gap is filled by a patch of complementary nucleotide sequence removed from the sister DNA duplex (Figure 6.10) which is 'pasted' into place by recombination-type processes involving RecA and other recombination proteins. The gaps left in parental DNA is filled by repair synthesis involving DNA polymerase and DNA ligase.

6.8 *Double-strand Break Repair*

In lower eukaryotes such as yeasts, double-strand breaks induced, for example, by ionizing radiation are primarily repaired by homologous recombination repair mechanisms similar to those described in the previous section. In these, DNA sequence information lost from the damaged chromosome is retrieved by exchange with its undamaged partner. Genetic analysis in *Saccharomyces cerevisiae* has established the involvement of a number of key proteins, for example Rad50 (a DNA-binding protein), Rad51 (a protein with considerable homology to the *E. coli* RecA protein), Rad52 (a protein that interacts with Rad51), Rad53 (an essential serine/threonine protein kinase) and Rad54 (a DNA helicase).

In higher eukaryotes there is now some evidence that double-strand breaks may also be repaired by homologous repair mechanisms, and homologues of Rad51 and Rad50 have been detected. Moreover, recent observations have revealed an association of the product of the breast cancer susceptibility gene *BRCA2* with human Rad51 protein, as well as a complex of involving human Rad50 with *nibrin*, a novel double-strand break repair protein, which is mutated in *Nijmegen breakage syndrome*

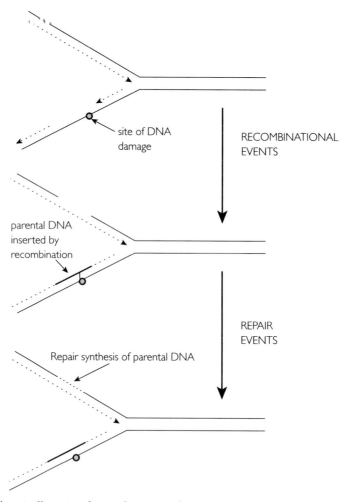

site of DNA
damage

RECOMBINATIONAL
EVENTS

parental DNA
inserted by
recombination

REPAIR
EVENTS

Repair synthesis of parental DNA

FIGURE 6.10 *A schematic illustration of post-replication recombination repair of a single-strand gap in DNA*

(see Chapter 8). However, there is also considerable evidence to indicate that repair can be carried out by mechanisms that do not rely on homology between the DNA molecules to be joined. Several components are required, but one of these, *DNA-dependent protein kinase (DNA-PK)*, activated by DNA ends, appears to be a crucial ingredient of the DNA double-strand break repair system. DNA-PK is a multiprotein complex that is only active when bound to DNA. It comprises a catalytic subunit (DNA-PK$_{CS}$) and a DNA targeting subunit Ku, itself made up of two subunits (Ku80 and Ku70). It is suggested that

DNA-PK may sense DNA damage and, by associating with DNA ends, acts to bring about the correct alignment of the broken ends and prevent the action of inappropriate nucleases. Additionally, it has been speculated that it might interact with DNA repair components and bring them to the site of DNA damage. DNA-PK can also phosphorylate and inactivate the transcriptional apparatus in the vicinity of double-strand breaks, thus possibly eliminating the complication of transcription during the repair of such breaks. However, since homologous DNA does not act as a source of nucleotide sequence information to guide this process, nucleotides may be lost and the ends of non-homologous DNA molecules may possibly be joined (illegitimate recombination), resulting in extensive chromosomal rearrangements.

6.9 *Repair of Damage to Mitochondrial DNA*

Because most reactive oxygen species that are generated within eukaryotic cells arise from mitochondria (see Chapter 4), it is perhaps not surprising that mitochondrial DNA (mtDNA) is found to have a higher steady-state amount of oxidative damage than nuclear chromosomal DNA. At present relatively little is known about DNA repair systems within mitochondria. Although mitochondria do not appear to have a nucleotide excision repair system and UV-photodimers in mtDNA are not removed, other damage to mtDNA, such as the oxidation of certain bases and strand breaks, can be repaired. The latter ability may reflect the presence of recombinational-type activities observed within mitochondria.

6.10 *Literature Sources and Additional Reading*

BOULTON, S.J. and JACKSON, S.P., 1996, *Saccharomyces cerevisiae* Ku70 potentiates illegitimate DNA double-strand break repair and serves as a barrier to error prone DNA repair, *EMBO Journal*, 15, 5093–103.

BRITT, A.B., 1996, DNA damage and repair in plants, *Annual Review of Biochemistry*, 65, 75–100.

CARNEY, J.P., MASER, R.S., OLIVARES, H., DAVIES, E.M., LE BEAU, M., YATES III, J.R., HAYS, L., MORGAN, W.F. and PETRINI, J.H.J., 1998, The hMre11/hRad50 protein complex and Nijmegen breakage syndrome: linkage of double-strand break repair to the cellular DNA damage response, *Cell*, 93, 477–86.

CROTEAU, D.K. and BOHR, V.A., 1997, Repair of oxidative damage to nuclear and mitochondrial DNA in mammalian cells, *Journal of Biological Chemistry*, 272, 25409–12.

FRIEDBERG, E.C., 1996, Relationships between DNA repair and transcription, *Annual Review of Biochemistry*, 65, 15–42.

FRIEDBERG, E.C., WALKER, G.C. and SIEDE, W., 1995, *DNA Repair and Mutagenesis*, Washington, DC: ASM Press.

HENLE, E.S. and LINN, S., 1997, Formation, prevention and repair of DNA damage by iron/hydrogen peroxide, *Journal of Biological Chemistry*, 272, 19095–98.

JACKSON, S.P., 1996, The recognition of DNA damage, *Current Opinion in Genetics and Development*, 6, 19–25.

JACKSON, S.P. and JEGGO, P.A., 1995, DNA double-strand break repair and V(D)J recombination; involvement of DNA-PK, *Trends in Biochemical Sciences*, 237, 412–15.

KOLODNER, R.D., 1995, Mismatch repair: mechanisms and relationship with cancer susceptibility, *Trends in Biochemical Sciences*, 237, 397–401.

LEHMANN, A.R., 1995, Nucleotide excision repair and the link with transcription, *Trends in Biochemical Sciences*, 237, 402–5.

LIANG, F., HAN, M., ROMANIEKO, P.J. and JASIN, M., 1998, Homology-directed repair is a major double-strand break repair pathway in mammalian cells, *Proceedings of the National Academy of Sciences USA*, 95, 5172–77.

MODRICH, P., 1997, Strand-specific mismatch repair in mammalian cells, *Journal of Biological Chemistry*, 272, 24727–30.

NAEGLI, H., 1995, Mechanisms of DNA damage recognition in mammalian nucleotide excision repair, *FASEB Journal*, 9, 1043–50.

RAMOTAR, D. and DEMPLE, B., 1993, Enzymes that repair oxidative damage to DNA, in Halliwell, B. and Aruoma, O.I. (Eds) *DNA and Free Radicals*, New York: Ellis Horwood.

SEEBERG, E., EIDE, L. and BJORAS, M., 1995, The base excision pathway, *Trends in Biochemical Sciences*, **237**, 391–97.

SHINIGAWA, H., 1996, SOS response is an adaptive resonse to DNA damage in prokaryotes, in Fiege, U., Morimoto, R.I., Yahara, I., and Polla, B. (Eds), *Stress-Inducible Cellular Responses*, pp. 221–35, Basel: Birkhauser Verlag.

VOLKNER, E. and KARMITZ, L.M., 1999, Human homologues of *Saccharomyces pombe* Rad1, Hus1 and Rad9 form a DNA responsive protein complex, *Journal of Biological Chemistry*, **274**, 567–70.

WALKER, G.C., 1995, SOS-regulated proteins in translesional DNA synthesis and mutagenesis, *Trends in Biochemical Sciences*, **237**, 416–20.

WEAVER, D.T., 1995, What to do at an end: DNA double-strand break repair, *Trends in Genetics*, **11**, 388–92.

WOOD, R.D., 1996, DNA repair in eukaryotes, *Annual Review of Biochemistry*, **65**, 135–67.

ZHANG, H., TOMBLINE, G. and WEBER, B.L., 1998, BRCA1, BRCA2 and DNA damage response: collision or collusion? *Cell*, **92**, 433–36.

CHAPTER 7

Genetic Information Flow

- Gene expression and its regulation provide the connection between *genotype* and *phenotype*.

- The chapter summarized the biochemical nature of genes and the basic mechanisms of transcription and translation that govern their expression.

- The effect of environmental factors on the expression of bacterial genes is primarily exerted through the control of transcription.

- Control over gene expression in eukaryotes is more complex than in bacteria, occurring not just at transcription but also at a number of other stages.

- Special non-coding DNA 'control' sequences are vital for the regulation of gene expression in response to environmental and other factors. Such sequences include promoters, enhancers, terminators, intron–exon splice sites, polyA addition sites, ribosome binding sites, and translation start and stop codons.

This chapter aims to provide a very broad general summary of current knowledge regarding the nature of genes and the mechanisms that control their expression, in both prokaryotes and eukaryotes. This will provide a basis for an appreciation of the consequences of gene damage brought about by environmental factors in terms of mutation and gene expression (Chapter 8), as well as the mechanisms that regulate gene activity in response to specific environmental stress (Chapter 9).

Written in the nucleotide sequence of an organism's DNA are the basic instructions for building that organism, often referred to as the *genotype*. However, as pointed out in Chapter 1, the *phenotype* of a particular organism is a product of these genetic instructions combined with environmental influences. The DNA inherited by an organism leads to specific characteristics by prescribing the precise construction of certain proteins. There is a flow of information from the genes of an organism into the construction of specific proteins, generally referred to as *gene expression*. It is the spectrum of proteins produced that provides the connection between genotype and phenotype. Critical to an understanding of the impact of environmental factors on the phenotype is an appreciation of the molecular processes that contribute to gene expression and its regulation.

The two basic mechanistic processes in the flow of genetic information within an organism are *transcription* and *translation*. Proteins are not constructed directly from genes. The link between genetic information in DNA and protein construction is *RNA (ribonucleic acid)*. Like DNA, RNA is a polymer of four nucleotide building blocks. However, in contrast to DNA, the bases in the RNA nucleotides are adenine (A), guanine (G), cytosine (C) and uridine (U) instead of thymine, and the sugar component is ribose rather than deoxyribose.

Transcription is the process whereby the assembly of RNA is specifically directed by the DNA of the genes. Translation is the processes by which the creation of protein molecules is dictated within cells by the RNA molecules emanating from the transcription process.

7.1 *Transcription*

In transcription, the sequence of nucleotides in DNA provides a template for assembling a unique sequence of RNA nucleotides. It is a polymerization reaction in which individual nucleotides are linked into a chain. The reaction is catalyzed by enzymes termed *RNA polymerases* which require a DNA template and the four ribonucleotide 5′-triphosphates, ATP, UTP, GTP and CTP, as substrates. The principles of the transcriptional process are shown in Figure 7.1.

The mechanism is initiated when an RNA polymerase molecule binds to the helical DNA molecule which caused its unwinding in the vicinity. One of the unwound DNA strands then acts as template. According to the rules of DNA–RNA complementary base-pairing (i.e. A pairs with U, and G pairs with C), the presence of guanine at this site causes the RNA polymerase to bind CTP from the cellular pool of the four ribonucleoside 5′-triphosphates. Subsequent to binding of CTP, the RNA polymerase undergoes a conformational change, causing it to respond to the next base in the template strand, which is adenine. This now induces the enzyme to bind the complementary UTP. This UTP is then liked enzymatically to the previously positioned CTP with the formation of a 3′–5′ phosphodiester bridge with release of pyrophosphate. In response to the formation of this first phosphodiester bridge, the polymerase moves

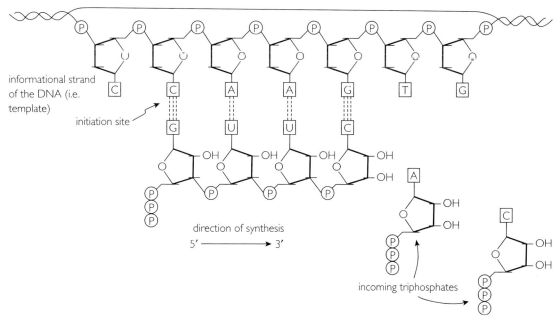

FIGURE 7.1 *The biosynthesis of RNA on a strand of DNA acting as template. Only the strand of DNA being transcribed (the information strand) is drawn out in detail. Synthesis proceeds in the 5′ to 3′ direction and the new RNA strand is 'antiparallel' to, and complementary in nucleotide composition to, the information on the template strand*

along the DNA template to the next DNA base, which in Figure 7.1 is another adenine. This specifies the insertion of another uracil nucleotide, which is linked through the second phosphodiester bridge to the new RNA chain. The next DNA base is a guanine and, by the rules of base-pairing, this will specify the incorporation of a cytosine nucleotide in the growing RNA molecule (Figure 7.1). The process continues to yield an RNA molecule with a sequence of nucleotides precisely complementary to the sequence of nucleotides in the DNA template strand.

Within cells the overall process of transcription occurs in three basic phases – initiation, elongation and termination (Figure 7.2). The first of these, *initiation*, involves the attachment of the RNA polymerase to the DNA and binding by the enzyme of the first nucleotide to be inserted in the RNA molecule. The RNA polymerases bind to specific sequences in DNA called 'promoters'. Usually factors in

addition to the RNA polymerase are required for successful initiation. The auxiliary initiation factors in eukaryotic cells that bind initially to the DNA 'promoter' sequence form a structure which then fits a part of the RNA polymerase enzyme, thus facilitating the formation of a strong complex between initiation factors, RNA polymerase and a length of DNA corresponding to around 75 base-pairs. It is this complex that effects the localized destabilization and unwinding of about 17 base-pairs of the DNA helix.

Elongation then follows in which RNA nucleotides are added sequentially according to the nucleotide sequence present in the DNA template. The process exhibits almost perfect fidelity. The conformational changes that occur in the RNA polymerase in response to the template DNA nucleotide bases are such that only the correct complementary RNA nucleotides can be incorporated into the RNA chains as they are elongated. During

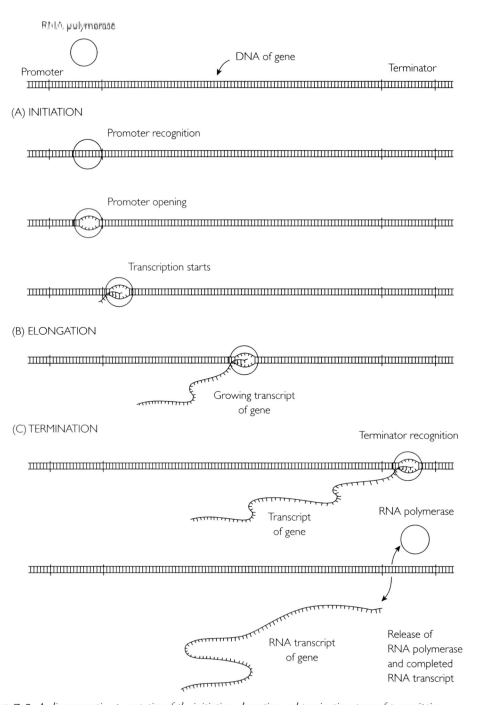

FIGURE 7.2 *A diagrammatic representation of the initiation, elongation and termination stages of transcription*

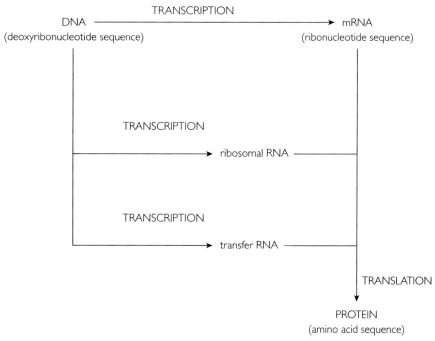

FIGURE 7.3 *The products of transcription and their cellular role*

elongation the DNA unwinds sequentially just ahead of the precessing RNA polymerase and rewinds behind it. This may involve auxiliary DNA topoisomerase-type enzymes.

Termination is the process by which the polymerization of RNA nucleotides is halted and the RNA product is released from the template. In prokaryotes, and possibly also eukaryotes, there are specific sequences in the DNA that act to signal the end of transcription. Termination factors such as ρ (rho), in conjunction with the RNA polymerase at these sites, act to destabilize the transcription machinery and release the RNA products.

In summary, during the transcription process deoxyribonucleotide DNA sequences serve as templates and are reproduced, but in terms of the ribonucleotide sequences of RNA. Certain DNA sequences when transcribed give rise to *transfer RNA (tRNA)* and others to *ribosomal RNA (rRNA)*. Transcription of yet other special sequences yield *messenger RNAs*

(mRNAs) which contain ribonucleotide sequences that will ultimately be 'translated' into the amino acid sequences of the protein vital to cell function (Figure 7.3).

7.2 *Translation*

The messenger RNA, together with transfer RNA and ribosomal RNA molecules, interacts within the cytoplasm of both prokaryotes and eukaryotes in the process of translation which results in the manufacture of proteins. It is termed 'translation' because during the assembly of individual proteins the genetic code, represented by a particular sequence of nucleotides in a messenger RNA molecule, dictates the specific translation into a particular sequence of amino acids in a protein. During this process, *ribosomes* read along a messenger

TABLE 7.1 *The genetic code showing the codons present in messenger RNA that specify individual amino acids in the construction of proteins*

Alanine	Asparagine	Aspartate	Arginine	Cysteine
GCG GCA GCC GCU	AAC AAU	GAC GAU	AGG AGA CGG CGA CGC CGU	UGC UGU
Glutamine	**Glutamate**	**Glycine**	**Histidine**	**Isoleucine**
CAG CAA	GAG GAA	GGG GGA GGC GGU	CAC CAU	AUA AUC AUU
Leucine	**Lysine**	**Methionine**	**Phenylalanine**	**Proline**
CUG CUA CUC CUU UUG UUA	AAA AAG	AUG	UUU UUC	CCC CCG CCA CCU
Serine	**Threonine**	**Tryptophan**	**Tyrosine**	**Valine**
UCG UCA UCC UCU AGC AGU	ACA ACG ACC ACU	UGG	UAC UAU	GUA GUG GUC GUU

RNA molecule gradually assembling a protein molecule with an amino acid sequence corresponding to the sequence of nucleotides in the messenger RNA. The nucleotides comprise a *genetic code* and in this context are 'read' three at a time as *codons*; each codon dictates the particular amino acid to be added at a specific position in a growing amino acid chain that will eventually constitute a protein. The genetic code specifically refers to the different three nucleotide codons in the messenger RNA and the particular amino acids they specify (see Table 7.1).

The actual process of constructing proteins takes place on small cytoplasmic particles called *ribosomes* which are complicated assemblies of

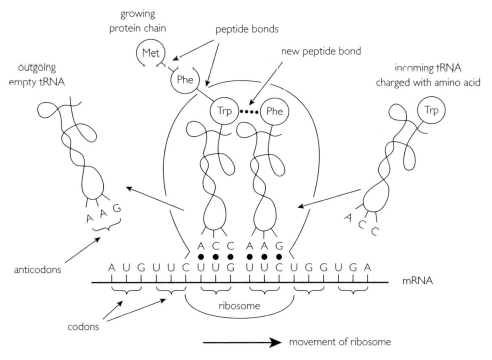

FIGURE 7.4 *A schematic diagram to illustrate the broad principles of peptide chain elongation*

ribosomal RNA and proteins that have many binding sites and catalytic centres. Essentially they can be viewed as enormous multisubunit enzymes with many active sites that recognize messenger RNA and other molecules and catalyze the reactions involved in progressing along the messenger RNA, reading one codon at a time and combining the designated amino acids to one another in the correct order at each step to create a specific protein (see Figure 7.4).

Initially each amino acid required for protein synthesis is covalently attached, in an enzyme-catalyzed reaction, to the transfer RNA molecule that contains its corresponding *anticodon.* This is a three nucleotide region in the transfer RNA molecule with the potential to form a set of three complementary base pairs with the *codon* in the messenger RNA for that particular amino acid.

In Figure 7.4 a hypothetical stretch of messenger RNA (mRNA) is illustrated with the nucleotide sequence *AUGUUCUGGUU-CUGGUGA* . . . During the initiation of protein synthesis, the mRNA attaches to the subunits of a ribosome. In the surrounding cytoplasm there is a pool of amino acids attached to their appropriate transfer RNAs. Among these, in this example, are methionine, phenylalanine and tryptophan. The anticodons of each transfer RNA can recognize and form complementary base-pairs only with the codons in the messenger RNA that correspond to the amino acids carried by the individual transfer RNAs. Thus at the ribosome, the anticodons of the transfer RNAs, carrying either methionine or phenylalanine, form complementary base pairs with the corresponding codons in the mRNA for methionine and phenylalanine, i.e. *AUG* and *UCC.* The amino acids methionine and phenylalanine are thus brought into close proximity by these interactions, and as a consequence of a nearby enzymic activity on the ribosome, the formation of a peptide bond between these particular amino acids is catalyzed. The incorporation of this initial methione is of general significance as the *initiation* of protein synthesis in all organisms always involves methionine and hence *the codon AUG is referred to as the 'initation codon'*. After this initiation step the ribosome then moves along the mRNA to the next codon, in this case *UGG* for tryptophan, and the process repeats, this time attaching a tryptophan to the growing protein. Once more the ribosome moves along to the next codon *UUC* in the mRNA, which in this case corresponds to the anticodon of phenylalanine transfer RNA, and subsequently the amino acid phenylalanine is attached to the growing protein chain. As each successive amino acid is added to the protein chain its empty transfer RNA is released. In this way translation proceeds, codon by codon, until the end of the coding region of an mRNA is reached. At this particular site instead of a normal type of codon the ribosome encounters one or other of three *'stop codons'* in the mRNA (*i.e. UAA, UAG, or UGA*). There are no transfer RNAs with anticodons that will base-pair with stop codons and the now-completed protein is released from the ribosome, and the ribosomes separate from the mRNA, in reactions catalyzed by specific release factors.

7.3 Gene Transcription and its Regulation in Bacteria

The preceding sections have presented a highly abbreviated overview of DNA transcription and mRNA translation. For more detailed descriptions of the molecular mechanisms relating to the different steps that comprise these processes, the reader is referred to more general advanced texts in molecular biology and biochemistry. An appreciation of gene expression and its regulation in bacteria is important for a general understanding of the environmental impact on genes and gene expression. In particular, the organization of bacterial genes into *operons* provides a very effective way in which bacterial gene expression can respond rapidly to changes in the environment. Critically the regulation of these gene responses involves use of particular non-coding DNA *'control sequences'* such as *promoters, terminators* and *operators*.

7.3.1 PROMOTERS, RNA POLYMERASES AND SIGMA FACTORS

Bacterial *RNA polymerase* comprises a 'core' structure of three different protein subunits, α, β and β', of 36, 150.6 and 155.6 kDa respectively. Two copies of the α-subunit are present, resulting in a structure of 380 kDa and the composition $\alpha_2\beta\beta'$. The β-subunits are structurally related to the large subunits of the eukaryotic RNA polymerases. The 'core' enzyme is capable of transcribing any DNA sequence into RNA. However, in order to initiate only specific transcription at promoter sequences in bacterial DNA, the core enzyme must first combine with a 70 kDa protein subunit designated *sigma* (σ^{70}) *factor*. The sigma factor has sites that will bind both to the 'core' RNA polymerase and to the particular DNA nucleotide sequences that constitute *promoters*. Such promoter sequences in bacterial DNA normally lie between 10 and 35 base pairs just upstream from the start of a gene (see Figure 7.5). The major form of RNA polymerase in *E. coli* that interacts with most promoters contains σ^{70}, but alternative sigma

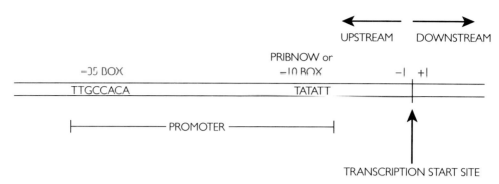

FIGURE 7.5 *The nucleotide numbering and general layout of a bacterial promoter region*

factors exist that enable the RNA polymerase 'core' to distinguish between different types of bacterial genes. This constitutes an important component of the mechanisms regulating transcription in bacteria, particularly in response to environmental changes. For instance, σ^{28} is involved in the recognition of genes involved in mobility and chemotaxis, σ^{32} for the recognition of genes regulated by heat stress, and σ^{38} for genes required in the stationary growth phase.

Two types of sequence can normally be found within the promoters of genes in *E. coli* that are specifically recognized by the combination of 'core' RNA polymerase and σ^{70}. One of these is usually centred about 10 base-pairs 'upstream' from the transcription start site, with a consensus sequence, *TATATT*, sometimes called the *Pribnow or −10-box*. The other has a consensus sequence, *TTGCCACA*, which usually occurs some 35 base-pairs 'upstream' from the start-site and is sometimes referred to as the *−35-box* (see Figure 7.5). (The form in which a consensus sequence is written indicates the nucleotides most often found at each particular position. It is the most typical form of the sequence.)

While the interaction of the polymerase complex with these promoter sequences essentially sets the basal transcription rate, this rate can be adjusted by two types of regulatory protein termed *repressors* and *activators*.

Repressors appear to exert their negative regulatory effect by binding close to the DNA promoter sequence and either directly inhibiting access by the RNA polymerase complex, or inducing conformational changes that have the same result. Activators can increase the rates of transcription by binding to recognition and control sequences overlapping or up from the −35 TTGCCA consensus sequence. Once bound, they appear to increase the strength of RNA polymerase binding by either reacting directly with the RNA polymerase or by inducing conformational changes in promoter DNA that increase its ability to bind the polymerase.

7.3.2 TERMINATORS

In bacteria the DNA nucleotide sequences signalling the end of transcription (*terminators*) come in two varieties. One requires the participation of a protein termination factor called rho (ρ) and another type causes the polymerase to cease transcription without the involvement of the ρ factor.

7.3.3 OPERONS

Although many bacterial proteins are present within cells in constant amounts, the

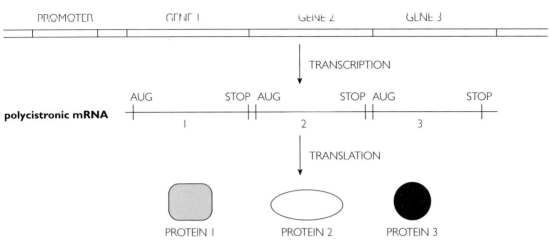

FIGURE 7.6 *The transcription of a bacterial operon into polycistronic messenger RNA*

concentrations of others are *environmentally* regulated, for example by the presence or absence of particular nutrients. In the main, bacteria have to adjust quickly to changes in their environment and they achieve this by regulating the production of proteins required for the breakdown and synthesis of a variety of compounds. Gene expression in bacteria is primarily controlled at the level of transcription. As bacteria have no nuclear membrane, mRNA synthesis and its translation are not physically separate, and occur simultaneously. Moreover, bacterial promoters often control the transcription of clusters of genes encoding proteins involved in related tasks. Such collections of related genes are referred to as *operons*, and are transcribed to give a single *polycistronic mRNA*. As can be seen from Figure 7.6, translation of this particular type of mRNA yields the different proteins encoded by the related genes. This is because there are translational start and stop codons along the length of the polycistronic mRNA to specify the particular regions to be translated.

Overall, the organization of bacterial genes into operons provides a very effective means whereby bacteria can respond extremely rapidly to changes in the environment, and ensures

that the enzymes required to metabolize a particular compound are made available simultaneously. Each operon has normally only one promoter site at which bacterial RNA polymerase binds prior to initiating transcription. The effectiveness of the promoter is controlled in turn by a series of regulatory components which respond to the bacterial environment and essentially modulate the affinity of RNA polymerase for the promoter positively, or negatively, and thus the rate of RNA chain initiation. As can be seen from the following examples, the regulation of many bacterial operons involves the interaction of small cellular molecules and specific regulatory proteins.

7.3.4 BACTERIAL OPERONS THAT CAN BE INDUCED OR REPRESSED BY ENVIRONMENTAL FACTORS

A well studied example of an *'inducible'* operon is the *lac operon* of E. coli (Figure 7.7). The human gut contains many *E. coli* bacteria that must respond to the sudden environmental occurrence, for example, of lactose. Although most of our food does not contain much lactose, milk contains relatively large amounts.

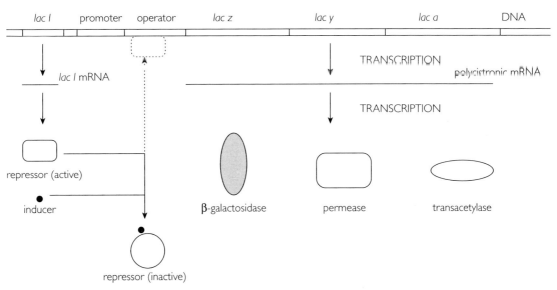

FIGURE 7.7 *The requirement of an inducer for the transcription of the* lac *operon*

The presence of lactose in the intestine very rapidly causes the resident *E. coli* to produce the enzyme *β-galactosidase* which cleaves the lactose to yield galactose and glucose. β-galactosidase is encoded by one of the genes, *lac z*, that make up the *lac* operon. The other genes that comprise the operon are *lac y*, which encodes *β-galactoside permease* (a carrier protein), and *lac a*, which codes for a *transacetylase*. This latter activity is believed to be important for the removal of compounds that are similar to lactose, but of no value to the bacteria. When no lactose is present, a *repressor* protein produced from the *lac i* gene binds to the *operator* nucleotide sequence adjacent to the promoter sequence, and thus interferes with the binding of RNA polymerase. This blocks the transcription of the *lac z*, *lac y* and *lac a* genes. On the other hand, if lactose concentration rises in the intestine following a drink of milk, it is converted to allollactose by the very small levels of β-galactosidase present in the *E. coli*. This allolactose acts as an *inducer*. It can bind readily to the repressor protein, which then undergoes

a conformational change such that it can no longer bind properly to the operator sequence. This allows the RNA polymerase to bind normally to the promoter and transcribe the three genes of the operon. By this means the bacteria can quickly adjust to the metabolism of lactose in the intestine. The concentration of lactose determines whether the necessary enzymes are synthesized, and the *lac* operon is said to be under 'negative' regulation by the repressor protein. Generally inducers of operon transcription are substrates of a catabolic pathway, or molecules closely related to its substrate.

An additional feature of the control of the *lac* operon is the involvement of the *catabolite activator protein (CAP)*. In situations where glucose is present in addition to lactose, priority to glucose utilization (which is energetically more favourable as it requires no additional enzyme synthesis) is ensured through the participation of an important small cellular control molecule called *cyclic adenosine monophosphate (cAMP)*. For optimal binding of RNA polymerase to the promoter sequence of

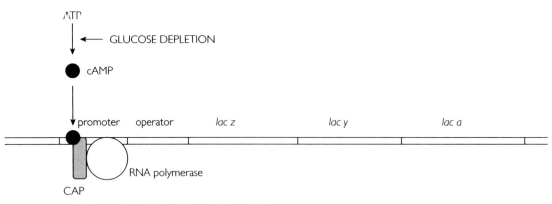

FIGURE 7.8 *Glucose, cyclic AMP and the regulation of the* lac *operon*

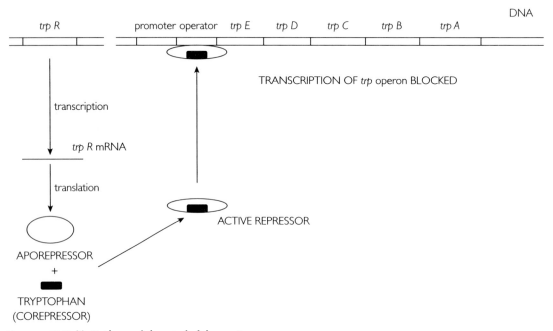

FIGURE 7.9 *Tryptophan and the control of the* trp *operon*

the *lac* operon, the formation of a complex between cAMP and the CAP is necessary. For example, when concentrations of glucose available to the bacteria are high, the intracellular levels of cAMP are low and therefore there is little formation of the cAMP–CAP complex. However, when glucose concentrations within the bacteria fall the extent of cAMP production from ATP rises, as does the

formation of the cAMP–CAP complex. This eventually is sufficient to facilitate the binding of RNA polymerase to the *lac* promoter, and so expedite the metabolism of lactose in place of glucose (Figure 7.8). Under such conditions the *lac* operon is under 'positive' regulation by the cAMP-CAP complex.

An example of a *'repressible'* operon is the *trp* operon (Figure 7.9). This operon comprises five genes responsible in bacteria for the synthesis of the amino acid tryptophan. The amount of tryptophan within a cell is regulated by blocking the transcription of the operon when there is sufficient tryptophan available. The gene *trp r* encodes an *aporepressor* which is only active and capable of binding to the *trp* operator sequence if complexed to tryptophan itself, which thus acts as *co-repressor*. Thus when the environmental concentration of tryptophan is high, this active repressor complex is formed and blocks the binding of RNA polymerase to the promoter, and thus prevents the synthesis of the enzymes required for tryptophan biosynthesis. In the absence of sufficient tryptophan, no active repressor complex can be formed, transcription of the *trp* operon can proceed and thus cellular supplies of tryptophan can be replenished.

7.4 *Eukaryotic Genes – their Structure and Transcription*

The RNA polymerases that catalyze the process of transcription of eukaryotic genes are large complex proteins (500–700 kDa). Each comprises two large protein subunits together with a number of smaller subunits (6–10). In bacteria, while only one type of RNA polymerase has the responsibility of transcribing all varieties of gene, eukaryotes have three basic types of polymerase. One of these, *RNA polymerase II*, transcribes genes that specifically encode the structure of proteins to yield messenger RNAs (mRNAs). In contrast, the other two eukaryotic enzymes, *RNA polymerase I* and *RNA polymerase III*, transcribe DNA sequences that give rise to ribosomal RNA (rRNA) and transfer RNA (tRNA), which, although non-coding, play key roles in the mechanics of protein manufacture (see Figure 7.3).

RNA polymerase II has at least 8–10 subunits ranging from 10 to 240 kDa. The largest subunit of around 230 kDa is very similar in all eukaryotes and also has some similarity to the largest subunit (β') of the bacterial RNA polymerases. The second largest subunit of eukaryotic RNA polymerase II is around 145 kDa and again there are structural similarities with the second largest subunits of RNA polymerases I and III. In addition, three of the smaller subunits, 27, 23 and 14.5 kDa, are common to all three eukaryotic polymerases. Two other small subunits, however, are only found in RNA polymerases I and II. In addition, each of RNA polymerases has a number of unique subunits as well as a variety of accessory factors, which play crucial roles in regulating the levels and specificity of transcription initiation at different promoter sequences on DNA. Some of these factors are fairly general and act in concert with all three of the polymerases, but others are highly gene-specific and dictate the particular gene, or group of genes, to be transcribed by different polymerases.

Transcription of the 'template' DNA strand of a gene commences at a nucleotide normally designated as +1 and proceeds in a *downstream* direction, the nucleotides in the transcribed DNA being given successive positive numbers. Downstream sequences are indicated diagramatically to the right of the *transcription start site (TCS)* in Figure 7.10. Conversely, nucleotides drawn diagramatically to the left are referred to as *upstream* sequences and are indicated by negative numbers. In eukaryotes, *promoter* sequences involved in the binding and activity of RNA polymerase lie upstream of the transcription start site of genes. In the case of genes whose transcription is regulated, promoters comprise two types of element. Those elements furthermost from the transcription start site, the *distal elements*, can be up to thousands of base-pairs from the

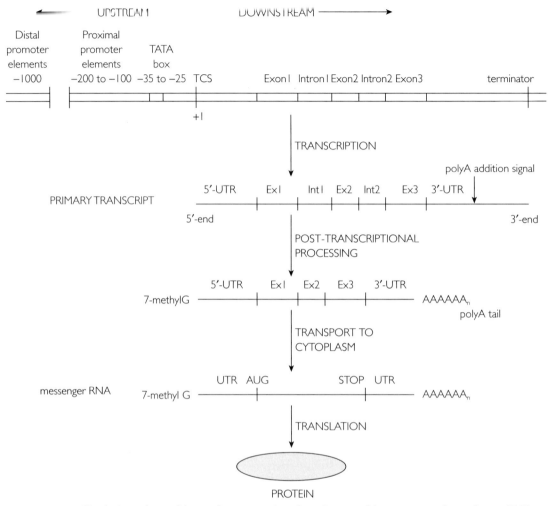

FIGURE 7.10 *Numbering and general layout of a representative eukaryotic gene and its upstream regulatory elements (TCS, transcription start site; Ex, exon; Int, intron; UTR, untranslated region; AUG, the translation initiation codon; STOP, any of the translation termination codons; 7-methylG, 5'cap structure; AAAAAA$_n$, polyA tail)*

transcription start site, whereas the *proximal elements* are usually within 100–200 base-pairs of the transcription start site. A number of studies indicate that the distal elements, which include which include sequences referred to as *enhancers*, regulate the extent of gene transcription, whereas the proximal elements serve not only as specific transcription start-site selectors, but also to mediate cell-specific gene transcription. A notable proximal element

lying some 25 to 35 bases upstream of the start site is the *TATA box*. Its role is to position RNA polymerase II for correct transcription initiation, and its name derives from its consensus sequence, *TATA(A/T)A*.

Although the rate of transcription of a large proportion of eukaryotic genes is subject to cellular regulation depending on various natural and environmental stimuli, there are genes that are continuously transcribed at a constant level. Sometimes these are referred to as the *housekeeping genes*, and although they have no TATA-boxes, their promoters are still recognized by RNA polymerase II. These promoters have an unusually high level of CG sequences −100 to 200 base-pairs upstream from the start site and are often dubbed *CpG islands*.

A striking feature of the coding regions of most eukaryotic genes is that they are interrupted by non-coding sequences. These are referred to as *introns*, and can be as much as ten times longer than the coding sequences. On average there are six introns per eukaryotic gene, and they are transcribed normally by the RNA polymerase II along with the adjoining coding regions, called *exons*, to yield a *primary transcript* (see Figure 7.10). This is essentially a precursor to the messenger RNA which will duly appear in the cytoplasm. As part a set of reactions − referred to as *post-transcriptional processing* − which occurs in eukaryotic cell nuclei, the RNA sequences, corresponding to the transcription of introns, are cut out of the primary transcript in a *splicing* process orchestrated by a special class of nuclear ribonucleoprotein particle called *spliceosomes*. Other components of post-transcriptional processing involve the addition of the modified base *7-methyl guanine* to *cap* the 5′-end of the messenger RNA, and the addition of a *tail* to the other end (3′) in the form of a string of up to around 200 adenine nucleotides (the so-called *polyA*). This is achieved by a dedicated nuclear enzyme called *polyA polymerase*, which in animal cells seems to take its cue from the occurrence of a *polyA addition signal sequence, AAUAA*, which usually occurs towards the 3′-end of the primary transcript. A highly schematic structure of a generalized eukaryotic protein-coding gene is presented in Figure 7.10, together with an indication of the post-transcriptional processing of a primary transcript.

7.5 *The Control of Gene Expression in Eukaryotes*

There is now a considerable body of evidence that particular genes are only expressed at elevevated levels within cells of eukaryotic organisms at various stages in embryonic development, and by cells in different tissues, or by cells exposed to distinct environmental and other stimuli. While gene expression refers to the processes involved in the production of a functional protein from its respective coding sequence in the DNA, the overall process can be regulated at a greater number of different stages in eukaryotes than in bacteria. For instance, expression is modulated by the complex transcriptional regulatory mechanisms, the physical availability of the gene sequence within the actual eukaryotic chromosome, the post-transcriptional processing of the primary gene transcript, and the kinetics of mRNA transport to the cytoplasm from the nucleus, as well as the control of translation and the intrinsic stability of the eukaryotic mRNAs themselves.

7.5.1 TRANSCRIPTIONAL REGULATORY MECHANISMS

A considerable variety of proteins called *transcription factors* exist within cells, whose role is to bind to specific non-coding regulatory DNA sequences and either stimulate, or inhibit, the transcription of adjacent genes. The situation is very complex. Single genes can be controlled at a number of distinct regulatory sites by a range of such transcription factors. Additionally, single transcription factors may exert effects at regulatory sites common to many diverse genes.

The regulatory DNA sequence closest to a large number of genes is the previously

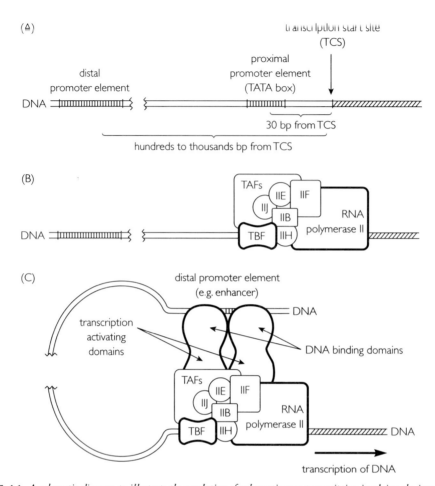

FIGURE 7.11 *A schematic diagram to illustrate the regulation of eukaryotic gene transcription involving the interaction of proximal and distal DNA promoter elements together with various transcriptional factors, TAFs: (A) the layout of DNA regulatory elements in relation to a hypothetical gene; (B) possible interactions of transcription factors, RNA polymerase II and proximal promoter elements; (C) a DNA loop model to account for the interaction of transcription factors bound to distal enhancer regions with factors interacting with RNA polymerase and proximal regulatory elements*

mentioned TATA-box. This is the location for the assembly of a group of transcription factors required for the facilitation of transcription of the gene by RNA polymerase II (Figure 7.11).

The establishment of such an assembly is sufficient in a number of instances to permit the transcription of the adjacent genes at a low basal level. It is the presence of other DNA-binding factors at other transcription regulatory sequences that can modify this basal rate upwards or downwards. Most transcription factors are proteins with two domains. Usually one domain has the role of recognizing and binding to the array of nucleotides that make up a specific DNA regulatory sequence. The other domain of the bound protein is then free to regulate the rate of transcription through interaction with other DNA-bound transcription factors (Figure 7.11).

7.5.2 THE ACCESSIBILITY OF EUKARYOTIC GENES WITHIN CHROMOSOMAL STRUCTURES

The accessibility of a gene to RNA polymerase II is influenced by the conformation of chromosome structures. Within each chromosome, the long linear DNA molecule is associated with various proteins to form *chromatin.* To summarize briefly, a primary feature of this structure is the packing of DNA in association with different types of the protein histone. Two molecules of each of the histones H2A, H2B, H3 and H4 make up octomeric core structures, around which are wrapped 146 base-pairs' worth of DNA. These assemblies, called *nucleosomes,* are linked to one another by approximately 60 base-pair stretches of naked DNA. This results in a 'beads-on-a-string' type of structure which, under certain conditions, can be condensed by further coiling to yield a thicker solenoid-like structure. Attachment of this structure at various points to a central protein-containing *scaffold* gives rise to looped chromatin domains corresponding to around 5000 to 200 000 base-pairs. The scaffold-attached structures can be further coiled during chromosome condensation to yield the highly compacted *mitotic chromosome* structures visible at *metaphase.* (For a more detailed description of the structural organization of eukaryotic chromatin and chromosomes, the reader is referred to the advanced texts in cell and molecular biology suggested in section 7.7.)

The precise relationship between the various types of chromatin structure and transcription has not yet been fully resolved. In general, the more condensed the chromatin structure, the less is the chance of transcription; although genes that are transcribed are in a relatively extended conformation, they still appear to contain nucleosomes. The precise location of nucleosomes within a transcription regulatory sequence may play a critical part in allowing DNA-binding transcription factors to influence transcription. For instance, a regulatory sequence could be accessible to a transcription factor because it lies between two nucleosomes. DNA sequences may play a role in such positioning of nucleosomes. Once bound, the transcription factor may bring about modification, or repositioning, of nucleosomes, to allow more ready access to other transcription factors, particularly the access of RNA polymerase II to the transcription start site. Present evidence suggests that, during the actual process of gene transcription, nucleosomes undergo minor structural modifications that allow the RNA polymerase/transcription factor complexes to progress along the DNA. There is a correlation between the transcriptional activity of chromatin and high levels of covalent modification of lysine residues of histone molecules by acetylation. Different acetylation patterns of histones, catalyzed by *histone acetyltransferases* and *deacetlyalases,* can help regulate the folding and unfolding of chromatin and its accessibility to the transcriptional apparatus.

7.5.3 POST-TRANSCRIPTIONAL PROCESSING OF THE PRIMARY GENE TRANSCRIPT

As part a set of reactions referred to as post-transcriptional processing that occurs in the nuclei of eukaryotic cells, the RNA sequences corresponding to the transcription of introns are cut out of the primary transcript in a 'splicing' process organized and controlled by a special class of nuclear ribonucleoprotein particles called spliceosomes. In the first step, cleavage of the primary transcript occurs at a specific nucleotide sequence at the 5'-end of the region corresponding to the transcribed intron, with the formation of a covalent attachment of the cut end to an internal adenosine residue, located within the intron region of the transcript. In the second step, the nucleotide sequence at the 3′-end of the region, corresponding to the intron, is then cut and the two coding regions, or exons, of the primary transcript are joined together covalently. Examination of the junctions between exons and introns from a wide variety of gene transcripts indicated the presence of a specific nucleotide sequence of ancient evolutionary origin. At the 5′-end of the introns, the most commonly found sequence is *G/GU*, and at the 3′-end the sequence is usually *AG/G*. Critically, a single nucleotide change within these evolutionarily conserved *splice-sites* is sufficient to block the removal of the intron from the primary transcript, which can accumulate in the unprocessed form. This is the case for individuals afflicted with β°-thalassaemia where the primary transcript of the β-globin gene is improperly processed.

In contrast, there appear to be no introns in bacterial genes. This eliminates the requirement for their removal from any primary transcript in bacteria. In addition, there is no requirement for a 5′-cap structure in bacterial messenger RNA.

7.5.4 NUCLEOCYTOPLASMIC TRANSPORT OF MRNA

In eukaryotic cells, the nucleus is separated from the cytoplasm by a double membrane system known as the *nuclear envelope*. As a consequence, the processes of transcription and translation are spatially segregated. This necessitates mechanisms for selective transport of RNAs and proteins in and out of nuclei and thus offers opportunities for gene expression regulation that do not exist in prokaryotes. Transport occurs through *nuclear pore complexes* (*NPCs*) in energy-dependent processes, the specific transport of different RNAs and proteins requiring partly distinct factors. NPCs are multiprotein cylindrical structures consisting of a basic framework (a spoke-type complex embracing a central channel), positioned between cytoplasmic and nuclear ring structures. Vertebrate NPCs have an outer diameter of 120 nm and are made up of around 1000 protein units (50–100 different types), known collectively as *nucleoporins.*

The signals that specify the transport of RNAs are generally believed to associate with RNA-binding proteins. Binding of such proteins appears to depend as much on RNA nucleotide sequence as on RNA secondary structure. In the particular case of mRNA, export signals appear to be associated with proteins that bind to the primary transcript. One of these, a protein in vertebrates called A1, shuttles rapidly between nucleus and cytoplasm.

7.5.5 TRANSLATIONAL CONTROL

Control of the translational processes in the cytoplasm can be brought about through the interaction between certain nucleotide sequences in mRNAs and cytoplasmic protein factors. Messenger RNAs have regions of either end which do not actually contain codons. These *untranslated regions (UTRs)* nevertheless contain nucleotide sequences important for the regulation of the overall translation process by the ribosomes. They can facilitate the selective binding of ribosome subunits to the specific regions of the mRNA that contain the AUG start site for translation (the codon for methionine). Additionally, the cytoplasmic location of a particular mRNA can be specified through interaction of UTRs with particular cytoskeleton proteins. This is important in processes relating to embryonic development.

Environmental stimuli such as heat shock, viral infection and iron toxicity can also influence the rate of translation. In the case of iron toxicity, control of translation of the mRNA for the iron-binding protein, ferritin, involves a protein interaction with the UTR at its 5′-end. In contrast, other situations are documented where the 3′-UTR is involved in interactions not with protein factors, but with novel regulatory RNAs that regulate the rate of mRNA translation.

Another potentially important aspect of translational control concerns the recently discovered phenomenon of *mRNA editing*. For example, in the case of the mRNA for apolipoprotein B, a cytidine at position 6666 is converted enzymically in intestinal cells to uridine. This results in the formation of a stop codon, *UAA*, at that position. Translation of this 'edited' mRNA thus gives rise to a shortened version of the protein, which is characteristic of cells of the small intestine. Messenger RNA editing also occurs within mitochondria and is extensive in many protozoa.

7.5.6 MRNA STABILITY

The control of gene expression can also be governed by the availability of mRNAs for translation. The mRNAs in bacteria are notably short-lived; in some cases degradation commences at their 3′-end even before synthesis of the RNA molecule has been completed. In contrast, most eukaryotic mRNAs have a relatively long lifespan, although the actual half-lives can vary considerably (from around 10 minutes to over 24 hours). Recent evidence indicates that the lifespan of an mRNA in a eukaryotic cell may be related to the length of its 3′-polyA tail. As already mentioned, during post-transcriptional processing, a tail of around 200 adenosine nucleotides is added to

the primary transcript. When the processed mRNA is finally exported to the cytoplasm, a number of *polyA binding proteins* are found associated with this tail. These particular proteins appear to sensitize the polyA tail to the activity of a special nuclease which progressively removes adenosine nucleotides, thus slowly reducing its overall length. The rate of polyA tail shortening is also influenced by the presence of specific nucleotide sequences within the 3′- untranslated region of the mRNA. However, once the length of the 3′-polyA tail is reduced to below 30 nucleotides, the overall stability of the mRNA is affected. Firstly its 5′-cap is removed, and then step-wise exonuclease degradation of the mRNA molecule proceeds from the uncapped 5′-end.

7.6 *DNA Control Sequences*

While certain regions of the DNA of an organism are dedicated, through the sequential arrangement of codons and the genetic code, to the construction of cellular proteins, it is clear that there are many other discrete nucleotide sequences which perform distinctly 'managerial' roles in regulating activities associated with genes. For instance, in preceding sections, the role of particular DNA (and RNA) nucleotide sequences was emphasized in influencing key processes such as the rate and extent of gene transcription (enhancers, promoters, etc.); mRNA splicing (intron–exon boundaries) and polyA addition (polyA addition sites); mRNA export to the cytoplasm; translational control (initiation and termination sequences); and mRNA degradation. The integrity of such control sequences is vital for the proper regulation of gene expression within eukaryotic cells. Any damage, or alteration, to such critical non-coding control sequences could thus have adverse consequences on cellular function, over and above any problems arising from alterations to coding sequences.

Since non-coding DNA makes up the bulk (85–90%) of the chromosomal DNA of a eukaryotic organism, considerable attention has focused on the possible functions of other non-coding DNA sequences. Specific non-coding regions of chromosomal DNA are implicated in organization of the overall process of chromosomal replication, as well as in the functioning of centromeres and telomeres. For instance, in yeast, sequences that confer the ability to replicate cellular DNA have been identified. These are referred to as *autonomously replicating sequences (ARSs)* and have an 11-base pair core sequence *ATT-TAT(A/G)TTTA. Centromeres* are important, since they acquire a complex protein structure at prophase, referred to as the kinetochore, which is essential for the attachment of the condensed chromosome to the microtubules of the mitotic spindle. The non-coding nucleotide sequences located at the *centromere region (CEN)* that are necessary for stable chromosome inheritance in yeast comprise

TCAT(...76–86 base pairs...)CCGAAA

Telomeres, which occur at the ends of linear nuclear chromosomes, also contain important non-coding DNA sequence elements. In humans and other vertebrates, this DNA sequence, *TEL*, comprises *TTAGGG* repeated around 1000 times and is associated with a special DNA replication process catalyzed by an enzyme, *telomere terminal transferase*. This enzyme is somewhat unusual inasmuch as it contains an RNA molecule that is complementary to the DNA strand of the chromosome that contains a 3′-end. This distinctive form of DNA replication is specifically required at the ends of linear DNA molecules, due the necessity for a double-stranded region to serve as 'primer' for new DNA synthesis in the 3′ to 5′ direction catalyzed by DNA polymerase. The special telomere DNA sequence is also capable of binding a special cellular protein that is thought to protect the chromosomal end from undesirable shortening that might result from random exonuclease digestion. Again, damage or alteration of the special DNA sequences at regions such as CEN, TEL and ARS is likely to have undesirable consequences for cells in terms of the overall organization, distribution, integrity and replication of eukaryotic chromosomes.

7.7 *Literature Sources and Additional Reading*

ALBERTS, B., BRAY, D., LEWIS, J., RAFF, M., ROBERTS, K. and WATSON, J.D., 1994, *Molecular Biology of the Cell*, New York: Garland Publishing.

BOLSOVER, S.R., HYAMS, J.S., JONES, S., SHEPHERD, E.A. and WHITE, H.A., 1997, *From Genes to Cells*, New York: Wiley-Liss.

FELSENFELD, G., 1996, Chromatin unfolds, *Cell*, **86**, 13–19.

GREGORY, P.D. and HORZ, W., 1998, Chromatin and transcription, how transcription factors battle with a repressive chromatin environment, *European Journal of Biochemistry*, **251**, 9–18.

GRIFFITHS, A.J.F., MILLER, J.H., SUSUKI, D.T., LEWONTIN, R.C. and GELBART, W.M., 1996, *An Introduction to Genetic Analysis*, 6th edn, New York: W.H. Freeman and Co.

GRUNSTEIN, M., 1997, Histone acetylation in chromatin structure and transcription, *Nature*, **389**, 349–52.

HOLSTEGE, F.C.P. and YOUNG, R.A., 1999, Transcription regulation – contending with complexity, *Proceedings of the National Academy of Sciences USA*, **91**, 2–4.

HUGHES, M.A., 1996, *Plant Molecular Genetics*, Harlow: Addison Wesley Longman.

KABLE, M.L., HEIDMANN, S. and STUART, K.D., 1997, RNA editing: getting U into RNA, *Trends in Biochemical Sciences*, **22**, 162–66.

KARP, G., 1996, *Cell and Molecular Biology – Concepts and Experiments*, New York: John Wiley and Sons.

LODISH, H., BALTIMORE, D., BERK, A., ZIPURSKY, S.L., MATSUDAIRA, P. and DARNELL, J., 1995, *Molecular Cell Biology*, 3rd edn, New York: Scientific American Books, W.H. Freeman.

MAYER, V.E. and YOUNG, R.A., 1998, RNA polymerase II holoenzymes and subcomplexes, *Journal of Biological Chemistry*, **273**, 27757–60.

NIGG, E.A., 1997, Nucleocytoplasmic transport: signals, mechanisms and regulation, *Nature*, **386**, 779–87.

RHODES, D., 1997, The nucleosome core all wrapped up, *Nature*, **389**, 231–33.

TJIAN, R., 1995, Molecular machines that control genes, *Scientific American*, February, 39–45.

WOLFE, S.L., 1993, *Molecular and Cellular Biology*, Belmont, CA: Wadsworth Publishing Co.

CHAPTER **8**

DNA Damage, Mutation, Cancer and Ageing

KEY POINTS

- Checkpoint controls arrest eukaryotic cells during the cell division cycle after damage to DNA, allowing repair to take place before errors are perpetuated as a result of the DNA replication process.

- Extensive DNA damage can lead to apoptosis (programmed cell death), thus protecting organisms from individual cells that are severely impaired.

- Less extensive DNA damage, or alteration, that escapes repair can be perpetuated by replication as a mutation.

- Analysis of the spectrum of location of mutational hot-spots may provide a means of establishing precise relationships between different environmental factors and specific genetic change.

- Humans with inherited diseases that manifest deficiencies in DNA repair are highly susceptible to the development of cancer.

- Evidence links mutations in protooncogenes and tumour suppressor genes with the development of cancers, and the possible role of environmental factors in the introduction of such harmful mutations is under investigation.

- Progressively unrepaired gene damage may also contribute to the process of ageing.

Recent developments in gene technology have provided us with the ability to target specific genes in the chromosomes of living organisms, thus permitting the construction of specific genetic deficiencies. For example, mice lacking the genetic information for certain types of DNA repair functions have now been produced. These manifest hypersensitivity to DNA-damaging agents, genetic instability, premature senescence, and elevated rates of cancer development, together with defective embryogenesis. Such abnormalities, displayed in DNA repair-deficient animals, attest to the crucial importance of DNA repair systems in protecting organisms from the effects of DNA damage.

8.1 Cellular Responses to DNA Damage

8.1.1 CELL DIVISION CYCLE CHECKPOINTS

In proliferating cells, an adequate response to DNA damage requires more than simply the repair of the lesion. Such damage is particularly threatening to cells during the processes of DNA replication and mitosis. Replication can render damage irreparable, while mitosis in the presence of unrepaired DNA strand breaks can lead to gross chromosomal aberrations in the progeny cells. To combat this menace, eukaryotes utilize mechanisms that temporarily arrest cell division processes at *cell division cycle checkpoints*. These are regulatory systems that control the order and timing of steps in the overall process of cell division, and ensure that crucial events such as DNA replication and chromosomal segregation at mitosis are completed with high fidelity. Cell division cycle

checkpoints respond, for example, to DNA damage by arresting the processes of cell division to provide time for the repair of DNA, and for the activation of genes that facilitate such repair. Checkpoint controls were first observed in the yeast, *Saccharomyces cerevisiae*, where it was shown that cells exposed to radiation were normally arrested before mitosis, whereas those with a mutation in the *rad9* gene were radiation-sensitive, did not arrest and subsequently died. This study defined certain checkpoint proteins as gene products that negatively regulated the cell division cycle in response to DNA damage. It is now clear, however, that the process of checkpoint control involves a network of interactions among many gene products.

Proliferating eukaryotic cells are characterized by a sequences of events leading to a duplication of their constituents. These events occur in a strict, temporal order. Two of the most obvious are the replication of the chromosomal DNA, the S-phase and the physical process of cell division, mitosis or M. Other phases are G1 and G2. The whole cell division cycle can be represented as a clock-face, as shown in Figure 8.1.

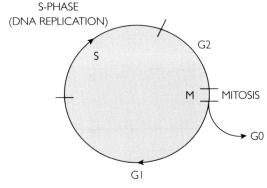

FIGURE 8.1 *The cell division cycle: S is the DNA replication, or synthetic, phase; G1 and G2 are the phases between mitosis (M) and S, and S and M respectively; G0 is the 'out of cycle' phase*

TABLE 8.1 *The association of different cyclin–CDK complexes with phases of the cell cycle*

Cyclin type	CDK type	Cell cycle phase
B	CDC2	Passage through M
D	CDK4 & CDK6	Transition into mid/late GI
E	CDK2	Transition to end of GI
A	CDK2	Transition between GI and S
A	CDC2	Passage through S

In practice, differences in the duration of G1 can be dramatic and account for the major variations in cell cycle time of different cell types. Moreover, cells that have spent a long time in G1 lose some of the enzymes concerned with DNA replication. Such 'quiescent' cells are often referred to as 'out of cycle', or in the G0 phase, and require a stimulus to urge them back into cycle.

The core components of the apparatus controlling passage of cells through this cycle are the *cyclin-dependent kinases (CDKs)* and their regulatory subunits, the *cyclins*. Cyclins activate the CDKs with which they associate and subsequently direct the activated kinase complexes to the appropriate cellular protein targets. It is the kinase-catalyzed phosphorylation that causes these proteins to perform their particular functions in the progression of cell cycle events. Although a complete understanding of the protein substrates phosphorylated by the different cyclin–CDK pairs is not available at present, some indications of the steps in the cell cycle progression that are mediated by individual cyclin–CDK complexes are given in Table 8.1.

While CDKs are present in cells at relatively constant levels, the amounts of their cyclin partners can vary considerably. This can be achieved through modulation of the transcription of their corresponding genes, or by modification of their intrinsic activity through phosphorylation/dephosphorylation reactions catalyzed by cdk-activating kinases and corresponding phosphatases. CDK inhibitors (CDKIs) also play important cell cycle regulatory roles.

Primarily, cell cycle checkpoints are inhibitory biochemical pathways. A particularly important example is the DNA-damage checkpoint. This detects damaged DNA and generates signal that arrests cells in the G1 or G2 phase, slows down the S phase and induces the transcription of DNA repair genes, thus avoiding the replication of damaged DNA in S phase. The precise position of arrest within the cell cycle, however, varies depending on the phase in which the damage is sensed, as genetic damage from exogenous or endogenous sources may be sustained at any point in the life of the cell. In animal cells many of the signals carrying information about DNA damage are channeled through a key cellular protein, referred to as p53.

The p53 protein, which can function as a type of cell cycle emergency brake, comprises 393 amino acid residues. Structurally the protein can be divided into several domains. The N-terminal domain is able to interact with transcription factors TFIID and TFIIH and thus act as an activator of transcription. This domain also has the potential to bind the MDM2 protein which, as will be outlined below, can act to repress the activity of p53. The central domain, between amino acid residues 100 to 300, has the ability to bind to specific DNA sequences. The C-terminal region, on the other hand, has a number of functions, including non-specific DNA binding, as well as being the region necessary for the oligomerization of p53 molecules. Within cells, p53 has a rapid turnover rate and is present at low levels in most tissues under normal conditions. However, certain types of cell stress such as DNA damage lead to increased cellular levels of p53. One possibility is that *stress-activated protein kinases*, or the *DNA-dependent protein kinase* described in section 6.8, modify p53 by phosphorylation, thus protecting it from degradation, and activating its function as a transcription factor for a number of genes which contain p53-specific upstream binding sites. One of the genes activated by p53 encodes the protein p21^{WAF-1} which induces arrest of the cell cycle, by acting as an inhibitor of cyclin D-CDK4/6, and blocks the cell cycle in G1. Another gene activated by p53 encodes GADD45, a protein which binds to the proliferation control nuclear antigen (PCNA) and thus also contributes to cell cycle arrest by inhibiting entry into the S phase. P53 also activates a gene for the protein MDM2, which plays a key role in the control of p53 itself. By binding to the N-terminal domain of p53, MDM2 not only blocks the ability of p53 to activate the transcription of the genes described above, but also targets p53 for proteolytic destruction by cellular proteosomes. Thus, p53 activation of the gene for MDM2 and the subsequent formation of a p53–MDM2 complex constitutes a negative feedback mechanism for maintaining low levels of p53 in normal cells as well as a means of terminating the p53 response to DNA damage (Figure 8.2).

Crucial questions remain regarding the precise nature of the surveillance mechanisms that detect DNA damage and induce the above inhibitions of cell cycle transitions. An attractive candidate is the *DNA-dependent protein kinase (DNA-PK$_{cs}$)*, which is known to be activated by DNA damage (see section 6.8). This protein belongs to a superfamily of eukaryotic *phosphatidyl inositol (PI) 3-kinases*, which also includes a protein termed *ATM* that is encoded by a gene mutated in patients suffering *ataxia–telangiectasia (AT)*. This is an autosomal recessive genetic disorder which is characterized by symptoms that include immune deficiencies, neuronal degeneration, growth retardation and a 100-fold increase in the incidence of cancers. Investigations have shown that cells from AT patients have high levels of DNA instability and are hypersensitive to ionizing radiation and other agents that bring about DNA strand breaks, suggesting that they cannot recognize, or repair, their DNA. In addition, these cells show reduced or delayed induction of p53 and are deficient with regard to the functioning of DNA damage-dependent checkpoints following radiation. It has been suggested that protein kinases in the ATM and DNA-PK$_{cs}$ subgroup of the PI-3-kinase family may function in pathways for sensing or responding to DNA damage in eukaryotes.

Bacteria also have systems that respond to DNA damage, although these serve a somewhat different purpose. In *E. coli*, DNA damage generates a signal that activates the RecA protein to destroy by proteolysis a repressor protein called LexA. This then permits the induction of a large battery of *'SOS response' genes*, some of which, as described in section 6.6, participate in *error-prone DNA repair* that forestalls cell death. Unlike the p53 responses in eukaryotes, the bacterial SOS response is very much a last resort and is essentially a trade-off between survival and the subsequent presence of mismatched nucleotides in bacterial DNA.

8.1.2 DNA DAMAGE AND APOPTOSIS (PROGRAMMED CELL DEATH)

Despite the checkpoint systems that signal cell cycle arrest in the face of repairable DNA

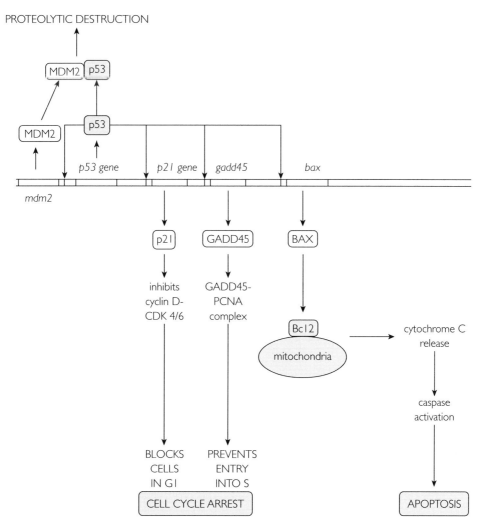

PROTEOLYTIC DESTRUCTION

FIGURE 8.2 *p53 and the activation of genes involved in cell cycle arrest and programmed cell death*

damage, it is now clear that more serious and potentially irreparable DNA damage may cause p53 and other regulatory molecules within cells to circumvent the cell cycle control systems entirely, and instead trigger *apoptosis* (sometimes referred to as programmed cell death). By this mechanism, tissues are purged of cells whose DNA, if replicated, might contain aberrations sufficient to cause the development of cancers.

Apoptosis has been recognized for some time as a physiological process of cell deletion that functions as an essential mechanism of normal tissue homeostasis, and which also plays a critical role in disorders such as cancer, AIDS and Alzheimer's disease. The health of multicellular organisms depends not only on their ability to produce new cells, but also on the ability of individual cells to self-destruct when they become superfluous or disordered. This process of cell suicide is integral to the building of tissues during development, and their functioning throughout adult life. Apoptosis is a deliberate and genetically programmed response by cells to certain normal developmental signals as well as to environmental stimuli. Overall, it is a subtly orchestrated disassembly of critical cellular macromolecular structures, permitting unwanted cells to vanish with the minimum disruption to surrounding tissues. It is characterized by nuclear and cytoplasmic condensation; fragmentation of chromosomal DNA at sites between nucleosomes by specific endonuclease action; collapse of mitochondrial membrane potential; and relocation of cytochrome c from mitochondria to the cytosol, where it helps to activate intracellular proteases called caspases. In contrast, 'necrosis' of cells is the outcome of severe injury and is not genetically controlled.

The fact that serious damage to cellular DNA can cause p53 to trigger apoptosis suggests that the cell division cycle apparatus and the apoptotic apparatus represent two distinct downstream targets of p53 signalling. At present, however, the mechanisms behind p53-induced apoptosis are only partially understood. Although p53 can activate the gene for p21^{WAF-1}, this cyclin–CDK does not appear to be involved in triggering apoptosis. p53 can nevertheless activate the transcription of genes encoding an apoptosis-promoting protein called Bax which can form a heterodimeric complex with another cellular protein, Bcl-2, to neutralize its activity (Figure 8.2). Within cells, Bcl-2 is associated with mitochondrial membranes and its normal role is to inhibit apoptosis, possibly by preventing the release of *cytochrome c* from mitochondria and thus preventing its activation of the caspase 'death' proteases. Release of the Bcl-2 safety catch and the loss of cytochrome c from mitochondria triggered by Bax also result in an elevated level of superoxide generation by the 'depleted' mitochondria. In terms of an overall mechanism for apoptosis, it now appears that, while stimuli can arise from cell nuclei in the form of DNA damage, stimuli can also arise from mitochondria and from the cell surface membranes. Ultimately, however, these different stimuli converge on the process of activation of *caspase enzymes*, which are cysteine proteases that drive most of the terminal events of apoptotic cell destruction.

Although cells from AT patients are defective in their ability to induce cell cycle arrest, as mentioned in the previous section, radiation-induced DNA damage in these cells can trigger p53-dependent apoptosis. It may be that the checkpoint arrest system is dependent on the ATM protein, whereas apoptosis pathways are not. It has been suggested that, at low levels of DNA damage, cellular levels of p53 are only transiently increased and the cell cycle checkpoint system only is engaged. Once the DNA damage is rectified, the levels of p53 return to their normal low value and the cell cycle resumes. In contrast, where there are high levels of DNA damage, the levels of p53 become continuously elevated, serving as a signal to trigger apoptosis.

As in animals, apoptosis plays important roles in embryonic, juvenile and adult phases of plant development. Although the morphological features differ a little from those observed in animal systems, apoptosis is also detectable during the responses of plants to attack by pathogens, where there is an extensive production of reactive oxygen species.

8.2 Mutation

Although most cells possess a wide range of detoxification and antioxidant defence systems, cellular DNA molecules nevertheless become damaged, depending on the severity of the exogenous or endogenous insult. As pointed out in Chapter 4, endogenous DNA damage is already high, and exogenous agents can produce an increment of lesions over this background level of endogenous lesions. DNA repair systems, however, are available to

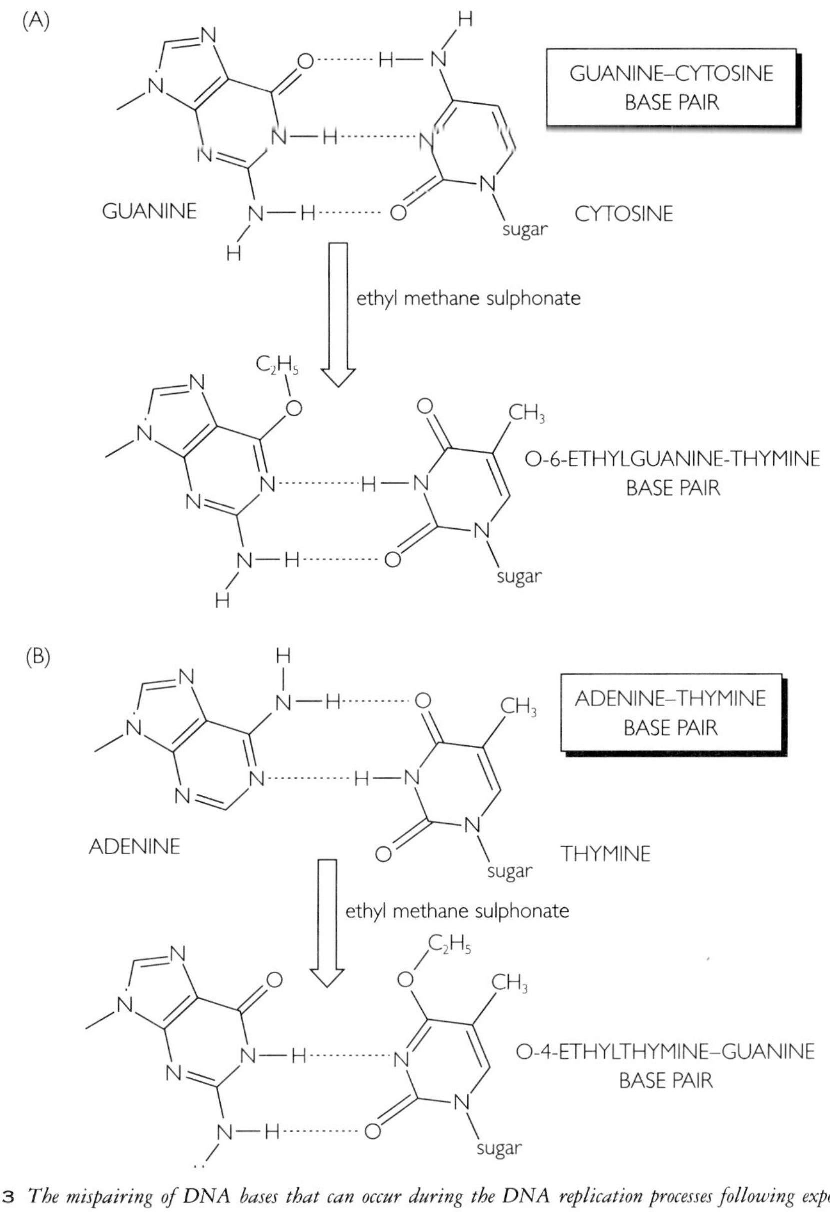

FIGURE 8.3 *The mispairing of DNA bases that can occur during the DNA replication processes following exposure of cells to the alkylating agent ethyl methane sulphonate: (A) the mispairing of 0-6-ethyl guanine with thymine rather than cytosine; (B) the mispairing of 0 4 ethyl thymine with guanine rather than adenine*

remove most (but not all) of the DNA lesions formed. The efficiency of the different repair systems towards the variety of possible lesions, such as base damage and strand breaks, is variable. Some of the lesions that remain are lethal because they block the replicative activities of DNA polymerases. Others have base mispairing properties and lead to mutations when cells replicate their DNA and divide. Mutations are heritable alterations in the nucleotide sequences of an organism's DNA. For example, in Figure 8.3, exposure of cells to the alkylating agent ethylmethane sulphonate brings about the modification of guanine residues to yield O-6-ethylguanine. If this base damage is not repaired, during ensuing replication it will mispair with thymine. The outcome is thus the change of a G–C pair to an A–T pair. Figure 8.3 also shows the mispairing that results from unrepaired O-4-ethylthymine residues.

Another type of mutational mechanism arises from the intercalation of dye molecules such as acridine orange, proflavine or ethidium bromide between DNA base-pairs (see Chapter 2.6). To accommodate this, the adjacent base-pairs move apart by a distance approximately equivalent to the normal distance that separates base-pairs. When the DNA strand containing the intercalated molecule is replicated the template and progeny strands can misalign, with the possibility that a nucleotide can be added or deleted in the progeny strand.

From Chapter 4 it will be appreciated the heritable alterations in nucleotide sequences can arise directly due to aberrations in the fidelity of the replication apparatus. A further example that has been the focus of much recent attention is the generation of variable numbers of end-to-end repeated copies of very short nucleotide sequences that arises from *replication slippage* (e.g. the prevalent repeated dinucleotide sequence CA that occurs in human DNA). In general, it is possible that repeated sequences provide the opportunity for slippage to occur. When the strand being replicated slips back on the template, further synthesis results in the insertion of extra bases, whereas when the template slips bases can be lost. Misalignment of DNA strands at replication may also contribute to the formation of larger scale insertions and deletions.

Other sources of spontaneous mutation, already outlined in Chapter 3, arise from the *tautomerization* of individual bases, the spontaneous *deamination* of bases and the hydrolytic loss of bases (e.g. *depurination*). The spontaneous deamination of an adenine base yields a hypoxanthine residue in its place. If this is not removed by base excision repair systems, during the subsequent DNA replication process it pairs with a cytosine nucleotide rather than a thymine, leading to the overall change of an A–T base pair to a G–C pair (see Figure 3.5). Spontaneous deamination of the minor base 5-methylcytosine gives rise to thymine. This in fact is a 'normal' DNA base, and therefore not subject to removal by repair systems, and pairs with adenine during the ensuing DNA replication process. Since the original 5-methyl cytosine normally pairs with guanosine, the outcome of this spontaneous change is the substitution of an A–T pair for a G–C pair (see Figure 3.6). Unmodified cytosine is also subject to spontaneous deamination, to yield uracil, normally found only in RNA. Uracil can, nevertheless, be removed from DNA by a DNA uracil glcosylase. If not, it will pair with an adenine rather than a guanine at the next round of DNA replication, thus substituting an A–T pair for the original G–C pair (see Table 8.2).

Spontaneous mutations can also arise from the activity of *transposable elements*. These are DNA sequences that occur naturally in the genomes of all organisms, but which are capable of readily altering their position within the genome. When present in non-essential locations of the genome they cause little harm, but transposition events can result in their translocation and insertion into other genetic loci, where the imposed change in nucleotide sequence may well have considerable genetic consequences. Similar outcomes can be induced by the insertion of *retrovirus* DNA equivalents into animal cell genomes (see Chapter 2), or the insertion into plant cell genomes of T-DNA sequences from the Ti plasmid of *Agrobacterium tumifaciens* (see Chapter 2).

TABLE 8.2 *The predominant inherited base pair changes brought about as a result of DNA alterations occurring spontaneously, or caused by environmental factors*

Environmental agent	DNA alteration	Base-pair change
	Deamination of 5-methyl cytosine to thymine	GC to AT transition
	Deamination of adenine to hypoxanthine	AT to GC transition
	Tautomerization of guanine to enol-form	GC to AT transition
Ethylmethane sulphonate	Alkylation of bases	GC to AT transition
	Alkylation of bases	TA to CG transition
Aflatoxin B$_1$	Adduct formation	GC to TA transversion
Ultraviolet radiation	Pyrimidine photodimer	GC to AT transition
Ionizing radiation	Formation of 8-oxo-guanine	GC to TA transversion
Oxidative stress	Formation of 8-oxo-guanine	GC to TA transversion
Benzo[a]pyrene	Adduct formation	GC to TA transversion
Acetylaminofluorene	Adduct formation	GC to TA transversion
Bisulphite	Deamination of cytosine to uracil	GC to AT transition

8.2.1 TYPES OF MUTATION

Mutations constitute a basic mechanism for genetic change. The simplest type of mutation is a *base substitution* in which a single nucleotide is replaced in the progeny DNA by a different nucleotide after replication. Some base substitutions replace one pyrimidine base with another, or one purine base with another. These are referred to as *transition mutations*. In other base substitutions, called *transversion mutations*, a pyrimidine is replaced with a purine or vice versa. Other types of mutation include *deletions* and *insertions*, as well as *translocations* and *repetitions* of DNA nucleotide sequences. *Mutagens* are agents that increase the frequency of all these types of mutational events.

Because the genetic code is read in groups of three nucleotides, or codons, there are potentially three 'frames' in which the mRNA can be translated. Only one of these results in the synthesis of the correct protein. A mutation that results in the insertion or deletion of a base will change the reading frame, and is called a *frameshift mutation*

AUG   AAG   UAU   AGU   ACU   C...

methionine   lysine   tyrosine   serine   threonine

If the U of the third codon (UAU in the above example) is deleted, the next available codon becomes AUA, which specifies the amino acid isoleucine. The 'new' reading frame next encounters the codons GUA and CUC respectively for valine and leucine

AUG   AAG   AUA   GUA   CUC...

methionine   lysine   isoleucine   valine   leucine

The substitution of a base to give a STOP codon (UAA, UAG or UGA) is known as a *nonsense mutation*. In the example below the U in the in the third codon specific for tyrosine has been changed to a G to yield the STOP codon UAG. On translation a truncated protein that will probably be non-functional will be produced

AUG   AAG   UAG   AGU   ACU   C...

methionine   lysine   STOP

A mutation that alters the codon from one amino acid to another is a *missense mutation*.

As shown below, the U in the forth codon which normally specifies serine has been substituted with an A, and as a result the codon now specifies arginine in place of serine. Such a change of amino acid will affect charge characteristics of the protein product and is likely to have deleterious effects on its functioning

AUG   AAG   UAU   AGA   ACU   c...
methionine   lysine   tyrosine   arginine   threonine

On the other hand, some missense mutations, by virtue the nature of the amino acids involved, may have little effect on the charge characteristics of the protein produced. Moreover, due to the nature of the genetic code some mutations may not alter the the amino acid specified by the codon (e.g. a change in the third base of the codon of the codon for valine will make no difference).

Although the above types of mutation relate to the coding sequences of DNA and RNA, a key point emphasized in the previous chapter is that many non-coding sequences have critical roles in the regulation of genes (e.g. promoters, enhancers, terminators, polyA addition sites, intron–exon splice-sites). Mutations in these 'control' sequences are also likely to have profound biological consequences.

Overall, a variety of different mechanisms can contribute to mutation and an impressive range of cellular repair systems function to eliminate many of the potentially mutagenic lesions in DNA. However, cells that are deficient in certain repair systems are observed to have higher rates of mutation. In terms of outcome, the phenotypic effects of spontaneous or induced mutations can range from minor cellular changes that can only be observed by sophisticated biochemical detection methods to extreme alterations in the behaviour of cells or organisms, and even death. Essentially the effect of a mutation is governed not just by the type of cell harbouring the altered version of the gene, but also by such considerations as the stage in the life cycle or development of the organism involved. Factors such as dominance or recessiveness of the mutant allele also influence the outcome in diploid organisms.

8.2.2 MUTATIONAL SPECTRA

With our developing ability to isolate and determine the complete nucleotide sequence of individual genes, combined with the use of *polymerase chain reaction (PCR)* techniques, gel electrophoresis and molecular hybridization methodologies (e.g. denaturing gradient gel electrophoresis, single-strand conformation polymorphism analysis, heteroduplex analysis, two-dimensional gene scanning, nucleotide primer extension), it is now possible to examine specific mutations and the process of mutagenesis in more detail. As it is important to know the type and the frequency of DNA adducts within cells, it is also necessary to understand their distribution along genes and the rate at which they are repaired. We also need to understand the types of mutation produced and which genes and gene positions are most prone to mutagenesis, either spontaneously or in response to specific environmental factors. Among other things, such knowledge is highly relevant to the problems of human genetic disease, as well as the development of cancers and the processes of ageing.

8.2.3 MUTATION RATE

The mutation rate is viewed as the probability that a gene undergoes a mutation in a single generation, or in the formation of a single gamete. The frequency of mutation can vary greatly from one gene to another. In the fruit fly, *Drosophila melanogaster*, the yellow body mutation occurs at 10^{-4} per gamete per generation. In contrast, the mutation in *E. coli* to streptomycin resistance occurs at 10^{-9} per generation. Moreover, within single organisms the frequency between individual genes can vary considerably. Early studies on eight different genes in corn showed the rates to vary at least 400-fold. In *E. coli* the frequency can vary between 10^{-5} and 10^{-9} per generation for some genes.

Spontaneous mutation has been observed to occur in certain DNA sequences at very significantly higher rates than is the case for other regions. These are often referred to as mutational 'hot spots', and usually contain runs of single nucleotides or short nucleotide sequences that are extensively repeated. Such sites may gain or lose nucleotides, possibly by mechanisms such as 'replication slippage'. Nucleotide sequences containing the modified base 5-methyl cytosine are also hot spots of mutation. As already discussed, the minor base 5-methyl cytosine can deaminate spontaneously to yield thymine, which will pair with adenine at a subsequent round of DNA replication. This results in the transition of a G–C base pair to an A–T pair (Table 8.2).

8.2.4 THE ENVIRONMENT AND RATES OF MUTATION

Since the 1930s it has been appreciated that certain environmental factors can significantly influence mutation rates. For example, at that time, it was found that sub-lethal exposure of bacteria and experimental animals to ionizing radiation can have dramatic effects. Several factors such as the level of environmental oxygen, however, can influence the frequency of mutations associated with a given dose of radiation. Fewer mutations per given dose are detected at low oxygen tensions. Microorganisms are also more likely to show higher rates of mutation under aerobic conditions.

Environmental temperatures can also affect mutation rates. Fruit flies raised at 27°C have two to three times the mutation frequency of flies raised at 17°C, and increased mutation rates can follow sudden fluctuations in temperature. Nutritional stress also appears to increase rates of mutation. When populations of bacteria are unable to grow because nutrients become exhausted, a subset of the population may start to mutate vigorously at random (hypermutation). Some of the resulting mutations may enable growth to resume. However, whether such hypermutation has evolved as a means of overcoming nutritional stress (which would have considerable evolutionary significance) or whether it is essentially a pathological response to damage accruing during stress remains to be determined.

8.2.5 OXIDATIVE DAMAGE AND MUTATION

Despite cellular antioxidant defences, chromosomal DNA can become oxidatively damaged. Oxidative damage is possible from reactive oxygen species generated as a result of exposure of cells to exogenous agents (e.g. ionizing radiation, cigarette smoke, asbestos), but also because of reactive oxygen species generated endogenously. DNA repair systems nevertheless excise most, *but not all*, of the oxidized bases. These are subsequently excreted in the urine and their measurement has led to estimates of the number of 'oxidative hits' on DNA that are actually repaired. This number is normally in the region of 100 000 per cell per day for the rat, and around 10 000 in humans. Despite such impressive rates of repair, there is considerable evidence for the accumulation of oxidative DNA damage, based on studies measuring the increase in 8-oxo-guanine in the DNA of humans and experimental animals as they age. Higher levels of 8-oxo-guanine, 8-hydroxyadenine and 2,6-diamino-4-hydroxy-5-formamidopyrimidine are also observed in the DNA of some cancerous tissues compared with the DNA of surrounding normal tissues. Increased levels of 8-oxo-guanine is also observed in the DNA extracted from lymphocytes of cigarette smokers as compared with non-smokers. In terms of mutagenic potential, 8-oxo-guanine nucleotides in DNA will pair with adenine nucleotides in addition to the normal cytidine nucleotides at replication. Overall this leads to the transversion of G–C base pairs to T–A pairs (Table 8.2).

8.2.6 TESTS FOR MUTAGENICITY

An early test for the ability of exogenous chemicals to induce mutations was the *Ames test* which utilized histidine requiring strains of *Salmonella* that were also deficient in nucleotide excision repair. These bacteria were mixed with a solubilized rat liver preparation (containing cytochrome P450 oxidases), exposed to the chemical under scrutiny and then plated on a growth medium containing no histidine. The test involved enumerating the bacterial colonies that were mutated in such a way that they were able to grow in the absence of histidine. By this means, chemicals that directly induced mutations in the bacteria were distinguished from those that required initial metabolism via P450s. While many types of bacterial mutagenicity tests have been developed over the years, their greatest use has been in classifying chemicals according to their biological activity, monitoring environmental samples and predicting possible toxicological effects.

Studies on the nucleotide sequence distribution of mutations in bacterial *lacI* gene induced by different chemical mutagens revealed mutational specificities that were characteristic of individual mutagens. For example, ethylmethane sulphonate, like ultraviolet light, caused the transition of G–C base pairs to A–T pairs, whereas aflatoxin B_1 brought about the transversion of G–C pairs to T–A. In the case of dipyrimidine lesions caused by ultraviolet light, an important recent observation is the preferential repair of lesions in the transcribed strand. This occurs in both bacterial and mammalian systems and reflects the coupling of transcription to nucleotide excision repair. Whether active genes are preferentially repaired by other cellular DNA repair systems is not yet known. Exposure to chemical mutagens has also revealed mutational specificities. The polycyclic hydrocarbon, benzo[*a*]pyrene, following conversion in animal cells to reactive diolepoxides by the action of cytochrome P450s, forms adducts with guanine. Typically such DNA adducts are higher in the tissues of tobacco smokers compared with non-smokers, and several types of study show that they give rise to the transversion of G–C base pairs to T–A pairs following replication of the modified DNA (Table 8.2). Similarly, exposure of cells to the aromatic amine, acetylaminofluorene, also produces guanine adducts and G-C to T-A transversions (Table 8.2).

8.2.7 MUTATION AT SPECIFIC GENETIC LOCI

Although individual mutagens exhibit a specificity with regard to the type mutation they promote (i.e. transition or transversion), the rates at which these mutations occur appears to be affected by the nature of the adjacent DNA sequences, certain DNA sites being more susceptible to the formation of lesions than others. This could also be a function of the physical accessibility of specific DNA sequences to the mutagens. In eukaryotes this is likely to be a function of individual chromosomal organization (e.g. the relationship with chromosomal proteins and extent of chromosomal condensation), or of the spatial geometry of chromatin within eukaryotic cell nuclei. Additionally, the mutation rate at a specific locus may reflect the rate at which lesions there can be repaired. As already pointed out, actively transcribing genes are known to be repaired preferentially by nucleotide excision repair systems, and rates of repair notably decline in ageing animals.

A number of recent studies have focused on mutations associated with specific mammalian genes. One particular gene that has received a great deal of attention is that which encodes the protein p53. This key mammalian cell regulatory protein, as already outlined, functions as a type of emergency brake on cell division control in the face of DNA damage. Mutations in the p53 gene are also of considerable relevance to cancer biology as they are implicated in the development of most of the major human cancers (see later in this chapter). 90% of the mutations in p53 genes, isolated from a variety of human tumour samples, are concentrated in the region of the gene that corresponds to exons 5 to 9. The whole gene is 20 000 base pairs and contains 11 exons. Mutational hot spots occur at codons 75, 248, 249 and 273 which encode the regions of the protein involved on DNA binding and stabilization. Codon 249 seems particularly prone to transversion mutation in human hepatocarcinomas, which occur with a high incidence in China, where aflatoxin B_1 is a notable risk factor.

Despite their intrinsic value, these studies on the spectrum of mutations affecting the p53 gene have been accumulated from the examination of DNAs isolated from pathological tumour samples. Unfortunately at present it is not possible to monitor continuously cigarette smokers *in vivo* for mutations in p53 genes of their lung epithelia. To survey the nature of mutations as they occur in living people, or other organisms, different strategies have been employed. In this context use has been made of transgenic animals into which various genes that code for readily measurable products have been introduced. These act as 'reporters' of mutational events within the cells of different tissues of the animal, following its exposure to specific exogenous mutagens. For example, transgenic mice, incorporating a bacterial *lacZ* gene sequence as a mutation 'target', have been particularly useful, as the mutated *lacZ* gene sequences can subsequently be recovered from isolated mouse DNA in lambda-bacteriophage vectors.

It has also become possible to measure *in vivo* mutation in human blood cells at certain genetic indicator loci. These may provide a pattern of mutation that may reflect mutagenic exposure. One useful indicator gene encodes an enzyme involved in purine metabolism called *hypoxanthine–guanine phosphoribosyltransferase (HPRT)*. For this purpose human peripheral blood is removed and T-cells carrying mutations in HPRT are cloned. The nucleotide sequences relevant to individual mutations can be characterized using the polymerase chain reaction (PCR). The typical mutation frequency of HPRT mutation in healthy adult is 5 to 8 × 10^{-6} per cell. However, the frequency of HPRT mutations not only increases with age of the subject, but is also elevated in the T-cells of tobacco smokers and in individuals undergoing exposure to radiation and chemotherapy. The gene has 9 exons and contains a coding region of 6567 base-pairs. Sequence information relating to a wide variety of mutations is accumulating, and in due course it will be possible to use mutations in the HPRT gene as a 'biomonitor' of human exposure to a wide variety of environmental mutagens. Predictable mutation patterns in specific genes may help to provide a means of identifying potential carcinogens.

8.3 *Human Diseases and Gene Repair*

A number of human hereditary diseases are now known either to be associated with defects in the cellular systems responsible for responding to DNA damage, or for correcting such damage (Table 8.3).

Such diseases have provided a powerful means of examining the mechanisms of DNA repair and repair responses in humans, as for many years the development of knowledge in this area was hampered by lack of suitable mutant cells. Another important feature of these hereditary diseases is that most sufferers have an abnormally raised incidence of cancer development. This has prompted the present impetus to explore the relationship between DNA damage, mutation and carcinogenesis.

The first human hereditary disease to be studied that had a clear deficiency in the ability to repair DNA was *Xeroderma pigmentosum*

(XP). Cells taken from victims are unable to carry out nucleotide-excision repair (NER) following exposure to ultraviolet radiation. At the clinical level XP is characterized by the extreme sensitivity to ultraviolet light of the exposed regions of skin, eyes and tongue. Patients also show a high probability of developing skin cancers as well as neurological disorders. Biochemical analyses have indicated that the repair deficiency leads to base mutations arising from (6-4)-photoproducts occurring predominantly in the transcribed DNA strand of active genes. The genetics of the disease is complex and the genes defective in XP fall into seven complementation groups. Mutations in genes *XPA*, *XPB*, *XPC*, *XPD*, *XPF* and *XPG* can all lead to defects in nucleotide-excision repair. Table 8.4 lists some of the possible biochemical functions ascribed to the products of some of these genes.

Trichothiodystrophy (TTD) is an autosomal recessive disorder characterized clinically by sulphur-deficient brittle hair and skin scales, as well as mental and physical retardation. Around half of the patients with TTD also report photosensitivity. Cell lines isolated from such victims exhibit nucleotide-excision repair deficiency of ultraviolet DNA damage. The repair defect appears to be linked in some way with a defect in the XPD gene product which is involved in the regulation of gene transcription (Table 8.4).

TABLE 8.3 *Human hereditary diseases due to defects in processing DNA damage*

Disease	DNA repair characteristics
Xeroderma pigmentosum	Defects in proteins of nucleotide excision repair
Fanconi's anaemia	Defective ability to repair DNA cross-links
Bloom's syndrome	Possibly defective DNA ligase
Ataxia-talengiectasia	Defect in response to DNA strand breaks
Nijmegen breakage syndrome	Defect in response to DNA double-strand breaks
Hereditary non-polyposis colon cancer	Defect in mismatch repair protein

TABLE 8.4 *Possible functions of* Xeroderma pigmentosum *gene products*

Gene product	Possible biochemical function in nucleotide excision repair
XPA	DNA binding (prefers UV-damaged DNA)
XPB	DNA helicase (part of transcription factor TFIIH)
XPC	Unknown
XPD	Part of transcription factor TFIIH
XPE	DNA binding
XPG	Damage-specific endonuclease

Cockayne syndrome (CS) victims are characteristically dwarfed, with other symptoms including photosensitivity. The condition is inherited as an autosomal recessive and, like TTD, may involve defects in the ability to transcribe certain genes. Following exposure of normal cells to damaging ultraviolet radiation, the nucleotide-excision repair system acts rapidly to remove photoproducts preferentially from strands of DNA that are being actively transcribed. This enables normal transcription processes to continue, but in cells from CS patients this does not seem to be the case.

Fanconi's anaemia (FA) is another autosomal recessive condition. Clinically it is complex, but primarily it is characterized by a depression in the level of all blood cells together with diverse congenital abnormalities. Moreover, FA victims have an increased risk of leukaemia and solid tumour development. In terms of DNA repair deficiency, cells from FA patients manifest an increased number of spontaneous chromosome aberrations, particularly in response to treatment with nitrogen mustards, mitomycin C and psoralens, that cause interstrand DNA cross-links. At the genetic level, at least four FA-related genes appear to exist. One of these, *FAC*, encodes a protein of around 63 kDa, which is cytoplasmic rather than nuclear. Its function is not yet clear. It is likely that *FA* genes may be involved in the control of haemopoietic development as well as in the repair of DNA cross-links and maintenance of DNA stability.

Ataxia-telangiectasia (AT) is a human autosomal recessive disorder which is characterized by symptoms that include immune deficiencies, neuronal degeneration, growth retardation and an approximately 100-fold increase in the incidence of cancers, particularly those of lymphoreticular origin. Cells taken from AT patients show increased levels of genomic instability and are hypersensitive to ionizing radiation and other agents that induce DNA strand breaks, suggesting that they are unable to recognize, or repair, such damage. In particular, they exhibit reduced or delayed induction of the protein p53 following irradiation, and are unable to respond normally at the DNA damage-induced cell division cycle checkpoints (see section 8.1). The gene *ATM*, which is mutated in AT individuals encodes a 350 kDa protein (ATM) which, as previously mentioned, has a carboxy-terminal region related to the catalytic domain of members of the phosphatidylinositol (PI) 3-kinase family. Recent suggestions are that the ATM protein either functions directly in DNA repair, or participates in a cell signalling pathway that modulates transcriptional and cell cycle control machinery in response to DNA damage. As outlined in section 8.1, there is a possible link between that activities of p53 and the ATM protein.

Bloom's syndrome (BS) is characterized by low birth weight and stunted growth together with the development of light-sensitive telangiectasia (broken vein appearance) on the facial skin. Around 10% of BS patients develop cancers at an average age of 20. Present evidence indicates that BS is inherited as an autosomal recessive and cells taken from BS individuals manifest a high mutation rate, abnormally slow DNA synthesis and chromosomal aberrations. Some evidence suggests that there may be a defect in a *DNA ligase* activity. However, although DNA replication intermediates accumulate in BS cells, these are longer than the normal sized Okazaki-fragments that might be expected if this were the only explanation. A defective DNA ligase may only be a consequence of some other primary gene defect in BS.

Nijmegen breakage syndrome (NBS) is a rare autosomal recessive disorder characterized by microcephaly, immunodeficiency and increased incidence of haemopoietic malignancy. In fact many of the clinical symptoms overlap with those of ataxia telangiectasia, although there are no neurological symptoms. At the biochemical level, features of NBS include ionizing radiation sensitivity and cell cycle checkpoint deficiencies. The gene that is defective in sufferers encodes a protein *nibrin*, or p95, which forms part of a complex, linking *recombinational DNA double-strand break repair* and cell cycle checkpoint functions.

Hereditary non-polyposis colon cancer (HNPCC) is one of the most common human cancer susceptibility syndromes known. Although colon cancers are the main types of cancer seen in HNPCC families, roughly 35–40% of HNPCC families develop other types of cancer and it now appears that HNPCC is associated with inherited defects in the *human mismatch repair system*. For instance, mutated versions of mismatch repair proteins hMSH2 and hMLH1 are prevalent in HNPCC patients (see also section 6.5).

8.4 Cancer Development

A significant characteristic shared by a number of human hereditary diseases associated with defective processing of DNA damage is the heightened susceptibility of the victims to the development of cancers. This is because an accumulation of somatic gene mutations can cause normal cells to become cancerous and life-threatening.

A normal healthy human cell responds to the dictates of its tissue environment. It only gives rise to progeny cells when the balance of stimulatory and inhibitory signals coming to it favour cell division. However, it is this very process of division, with its requirement to replicate chromosomal DNA, that (as explained above) carries the hazard of genetic mutations which could undermine the delicate regulatory networks controlling cell division. A single mutation may be sufficient to allow a cell to undergo the occasional unscheduled division cycle and be less responsive to the dictates of the extracellullar signals. If such mutations accumulate, progeny cells that are totally unresponsive to the normal external control factors may be generated. No longer will such cells respond to growth inhibitory signals; instead they gain the ability to grow uncontrollably and manifest signs of malignancy. The resulting mass of cells not only can impinge on, and damage, adjacent healthy tissue, but also can cross the barriers between one organ and another, and thereby 'metastasize' by setting up colonies at distant tissue sites.

8.4.1 PROTOONCOGENES AND ONCOGENES

Protooncogenes are a group of cellular genes whose mutation is central to the development

TABLE 8.5 *Some examples of human protooncogenes*

Protooncogene	Cellular function of gene product
	Receptors
erbB	Platelet-derived growth factor receptor
erbA	Thyroid hormone receptor
	Intracellular signalling proteins
ras	GTP-binding protein with GTPase activity
src	Protein kinase phosphorylating tyrosine
raf	Protein kinase phosphorylating serine/threonine
	Transcription factors
myc	Regulation of nuclear gene transcription
fos	Regulation of nuclear gene transcription

of tumours. They encode proteins that facilitate cell replication. For example, some encode receptors that protrude from the cell surface and bind serum proteins called growth factors. Interaction of growth factors with their appropriate cell surface receptor initiates an intracellular signal that ultimately leads to cell division. Other protooncogenes specify the amino acid sequence of proteins that function within cells to relay this growth signal, while still others encode transcription factors that regulate the expression of cell division control proteins (Table 8.5).

Mutation associated with protooncogenes converts them to potentially carcinogenic forms known as *oncogenes*, which can drive excessive cell multiplication. They are inherited as dominant mutations; just one mutated protooncogene allele is sufficient to initiate the deregulatory processes. The mutations may cause the protooncogene to produce too much of its growth regulatory protein, or to produce an altered form which is inappropriately active. For example, an aberrant growth factor receptor could release proliferative signals into the cell even when no growth factors are present. Alternatively, the protein encoded by an abnormal *ras* gene, for instance, could transmit signals intracellularly, even when

growth signals from receptor are not prompting such action. Indeed, such hyperactive Ras proteins are found in around 25% of all human tumours, including carcinomas of colon, pancreas and lung. Another example concerns transcription factors. Under normal circumstances cellular levels of transcription factors, such as Myc, which promote cell division, are only elevated in response to specific growth factor-initiated signals. However, in many types of cancer Myc levels are kept constantly elevated even in the absence of growth factors. Although individual oncogenes can initiate the deregulation of cell growth, by themselves they are unable to cause cancers. Other oncogenes are necessary, as is, usually, the participation of another class of genes called tumour suppressor genes.

8.4.2 TUMOUR SUPPRESSOR GENES

A number of inherited mutations predispose people to familial cancers that are not oncogenes. In addition to defects in genes encoding DNA repair functions, another important group of genes dictate the synthesis of cellular proteins that inhibit cell replication. They

TABLE 8.6 *Examples of some human tumour suppressor genes*

Supressor gene	Familial condition	Associated cancers
p53	Li-Fraumeni syndrome	50% of all human cancers
Rb	Retinoblastoma	Retinoblastoma; osteosarcoma; breast, lung and bladder carcinomas
APC	Adenomatous polyposis coli	Colon and brain cancers
DCC		Colon cancer
Nf-1	Neurofibromatosis type 1	Neurofibromatosis; colon cancer; astrocytoma
BRCA1	Breast and ovarian cancers	Breast and ovarian cancers

are inherited as recessives, and are referred to as *tumour suppressor genes* (Table 8.6).

Only the physical or functional loss of *both* alleles of tumour suppressor genes affects cell function. An example is the inactivation of both alleles of the *Rb* gene, which sets cells free of growth restraints and is involved in the development of retinoblastoma, as well as other cancers of the lung, breast and bladder. Another tumour supressor gene whose loss, or inactivation, removes restraints to proliferation is that encoding p53. To date, mutations in the p53 gene are found in half of all types of human cancers.

8.4.3 TUMOUR PROGRESSION

It is now clear that cancer development involves the accumulation in one cell and its direct descendants of mutations in both protooncogenes and tumour supressor genes. Initially a single cell suffers a mutation that allows it to divide under conditions that prevent normal cells from dividing. The next generation of cells derived from these inappropriately dividing cells carries the same mutation and shows the same aberrant growth. Some time later, one of these cells or their descendants undergoes a further mutation that further enhances its ability to evade normal regulation, possibly permitting it to pass

through normal tissue barriers and enter the blood. This additional mutation is then passed on to progeny cells. Repetition of this general sequence of events, referred to as *clonal evolution*, allows one original cell to accumulate the mutations necessary to *metastasize* and colonize other organs.

The development of colon cancers has been particularly well studied. One of the earliest mutations to occur in the development of this seems to occur in a gene called *APC* (adenomatous polyposis coli). Although its precise function is not yet clear, it would appear to function as a tumour suppressor gene, and, if both alleles of *APC* are damaged, then inhibition of DNA replication is relaxed. Colon cancers also carry mutated versions of two other suppressor genes, namely p53 and *DCC* (deleted in colorectal cancer), as well as a mutated *ras* gene. Although the *ras* mutation often occurs early in the progression towards malignancy, the order of occurrence of mutation in *ras*, *p53* and *DCC* is not fully established. The mutations in suppressor genes and protooncogenes relating to the development of many other tumours are currently being researched in many laboratories. As already mentioned, mutations in the supressor genes encoding p53 and RB protein occur in a high proportion of tumour types. In addition, mutations that were originally found in hereditary non-polyposis colon cancer (HNPCC) now appear to occur in about 0.5% of the human population and can dramatically increase the risk not only of colon cancer but also of ovarian, uterine and kidney cancers. The DNA involved normally encodes proteins involved in the process of mismatch repair (see section 8.3). Importantly, if DNA repair proteins are defective, then the number of mutations passed to daughter cells during the process tumour development may increase considerably. The progeny cells may then pass on DNA carrying even more mutations to their daughters. Defects in DNA repair genes may thus increase the speed at which certain tumours arise and become lethal.

8.4.4 ANEUPLOIDY AND CANCER

Although the major thrust of current thinking about the genetics of cancer centres around the somatic gene mutation hypothesis outlined above, it has been observed that the most impressive genetic abnormality of cancer cells is chromosome imbalance, or *aneuploidy*. A suggestion is that chemicals that act as carcinogens without apparently damaging DNA can cause aneuploidy and transformation of cells to the malignant state. They may achieve this by interfering with the DNA–protein interactions that are crucial in the relationships of chromosomes with the mitotic spindle apparatus. At present it is difficult to reconcile this view with the gene mutation hypothesis.

8.5 *The Environment and Cancer*

Despite the knowledge that has accumulated regarding the genetic basis of cancer development and the role of mutations, questions remain regarding precise causes of specific cancer and the role of environmental factors. The situation is complicated by the fact that the genes people inherit from their parents can also significantly influence cancer development. Not only can these include mutated forms of protooncogenes and tumour suppressor genes, but they will also include different variants of genes encoding DNA repair enzymes, cellular detoxifying enzymes and antioxidant defence enzymes. Variation in the activity of the enzymes encoded by this latter group of genes will clearly affect the efficiency with which cells can protect their genes from the deleterious effects of both exogenous and endogenous carcinogens. In general there appear to be two different categories of carcinogen, or cancer-causing agent. One type is able to 'initiate' by bringing about the alteration of specific genes involved in the regulation of cell replication and migration. The other category includes agents that do not actually damage genes but instead selectively 'promote' the growth of tumour cells, or their precursors.

For the past 50 years or so, despite these difficulties, epidemiologists have sought to evaluate environmental causes of cancer. However, epidemiologists can only identify the cause when people that have a specific type of cancer are regularly found to have a history of excessively high exposure to a particular factor.

8.5.1 OCCUPATIONAL HAZARDS

Over many years considerable knowledge has been accumulated, often at great cost in terms of human life, regarding the carcinogenic potential of a number of dangerous substances encountered in the workplace. Examples include asbestos, formaldehyde, diesel exhaust, benzene and radon. This knowledge now forms the basis of the legislative control of many such occupational carcinogens in the workplace. In the developed world at least, increasingly rigorous control measures over the past 50 years have halved the proportion of cancer deaths caused by occupational exposure.

8.5.2 RADIATION

In the 1930s, experiments with fruit flies showed that X-rays were powerful agents of mutation, and many early workers with X-rays subsequently died of cancer. It is now clear that ionizing radiation can produce single-strand DNA breaks, double-strand breaks and altered DNA bases. While single-strand breaks are normally repaired very efficiently, the other types of DNA damage are responsible for mutations and lethality. Cellular attempts to repair double-strand breaks often lead to extensive chromosomal translocations, inversions, duplications and deletions. After the nuclear accident at Chernobyl, where many people were heavily exposed to radiation and some increases in thyroid cancer in children were observed, studies in the populations as far as 200 km to the north revealed an approximate doubling of the normal somatic mutation rate even a decade later. This appeared to be due largely to an increased frequency of slippage events during DNA replication (see section 8.2). This aspect of DNA replication may be more sensitive to the effects of radiation than others. It is perhaps surprising, then, that among the survivors from the Japanese cities blasted by atomic bombs in 1945, and thus exposed to extremely high levels of ionizing radiation, only 1% actually died from cancers known to be related to radiation.

It has been argued from a number of early studies that if high-energy radiation can damage chromosomes and if these chromosomes are in sperm or eggs, then this damage should be inherited. Contrary to this expectation, studies of around 31 000 human descendants of these atomic bomb survivors showed little evidence for extensive hereditary effects. This dearth of mutations in later generations of humans may be the result of human sperm and eggs being protected in some way, or of extremely effective repair of germ cell DNA, or of highly active triggering of apoptotic death of germ cells carrying hopelessly damaged DNA. Despite these important statistics, there are still suggestions that the children of men exposed to radiation at work, for example at nuclear power stations, have a higher than normal risk of contracting leukaemia. However, close examination of some specific situations has indicated that the risk did not appear to be directly related to the parental external radiation dose, and it has been suggested that other factors, possibly leukaemia viruses, may be involved although there is no evidence of this as yet. Separate studies have shown that the risk of developing infant leukaemia following exposure, *within the uterus*, to radiation from the Chernobyl accident is approximately double, although no difference in leukaemia risk was seen in children aged 12 to 47 months. Moreover, irradiation *prior to conception* had no demonstrable effect on leukaemia risk in any of the age groups studied.

At present, only around 2% of cancer deaths in the USA can be specifically ascribed to any type of radiation. Of these, the major proportion is attributable to skin cancers caused by exposure to the ultraviolet rays of the sun. Three forms of skin cancer are observed: squamous cell carcinomas, basal cell carcinomas and melanomas. The last of these is relatively uncommon, representing only around 4% of annual skin cancer cases in the USA. Studies of squamous skin cell carcinomas, which occur especially on the faces and hands of whites living in the tropics, indicated mutations at particular sites in the tumour suppressor gene for p53. The same has been observed for basal skin cell carcinomas but not for melanomas. Dermatologists have recognized that, when skin is sunburned, some cells become apoptotic. In general, it seems that apoptosis, or programmed cell death, is important in the prevention of non-melanoma skin cancers. Loss of p53 function may prevent cells from initiating apoptotic processes in the face of high levels of DNA damage. In contrast, the mechanisms involved in the triggering of melanomas are not yet understood.

Other sources of radiation, such as electric power lines, household appliances and cellular telephones, do not contribute significantly, and diagnostic X-rays yield compensating benefits. Even the risks from radon, which has been linked to lung cancer, are now minimal with recent improvement in ventilation of buildings and mines.

8.5.3 ATMOSPHERIC ENVIRONMENTAL POLLUTANTS

At present tobacco smoke tops the lists of lethal carcinogens. In the USA it causes at least 30% of all cancer deaths. Cigarette smoking can cause many different cancers. Not only is it central to the generation of cancer of the lung, the upper respiratory tract, oesophagus, bladder and pancreas, but it also probably causes cancers of the stomach, liver and kidney as well as being implicated in chronic myelocytic leukaemia and colorectal cancer. Although the passive inhalation of environmental tobacco smoke causes less lung cancer than active smoking, as many as a few thousand deaths attributable to second-hand smoke occur annually in the USA.

Analyses of air samples from urban areas indicate that chronic exposure to high levels of air pollution can also increase lung cancer risk, particularly among those who already smoke. A major carcinogenic hazard is likely to be diesel exhaust, although sulphur dioxide from the burning of low-grade coal may also be important.

8.5.4 MEDICAL TREATMENTS

Considering the nature of some of the drugs designed to prevent propagation of cancer cells, it is perhaps not surprising that they that can damage DNA (see Chapter 2) and have themselves turned out to have carcinogenic effects. On balance, however, their clinical usefulness outweighs their risk. Certain immunosuppressive drugs can also be carcinogenic, as can some oral contraceptives and oestrogenic supplements taken to counteract menopausal effects.

8.5.5 VIRUSES

A number of viruses have now been implicated in the development of certain human cancers, e.g. papilloma viruses in cervical cancer, the Epstein-Barr virus in pharyngeal cancers and Hodgkin's disease, hepatitis B virus in hepatomas, and HIV (human immunodeficiency virus) in Kaposi's sarcoma. In addition, there is some evidence for human leukaemia viruses, and, in general, viruses may account for around 5% of cancer deaths in developed countries.

8.5.6 ALCOHOL

In cases where there is a large consumption of alcohol, especially by smokers, the risk of cancer in the respiratory tract and digestive tract increases. There is also increased risk of liver cancer, preceded by cirrhosis.

8.5.7 DIET

The proportion of cancer deaths due to diet in the USA has been suggested as equivalent to that caused by tobacco smoke. Overeating and lack of exercise also contribute to the risks of acquiring certain cancers. Obesity, for example, increases the risk of cancers of the colon, kidney and gall bladder. Some data suggests links between colon and rectal cancers, animal fat and red meat. Fat-rich diets are also believed to contribute to breast cancer.

In general, foodstuffs contain many thousands of different chemicals, both natural and synthetic, and it is extremely difficult to be specific with regard all possible carcinogens. Despite this, as pointed out in section 2.5, some food components such as nitrosamines and heterocyclic amines have well-established mutagenic effects and can act as carcinogens in experimental animals. Moreover, aflatoxins can cause liver cancer in humans. In general, although foodstuffs and beverages contain large numbers of compounds mutagenic towards bacteria, the ability of many of these compounds to cause cancer has only been demonstrated in experimental rodents, and only at concentrations vastly in excess of what might normally be encountered in the human diet.

Closer inspection of the possible connections between cancer deaths and diet has suggested that the absence of certain components may be crucial. For instance, diets deficient in fruits and vegetables can considerably increase the risks of a number of cancers. Epidemiological studies showed impressive protection by diets rich in fruits and vegetables against epithelial cancers of lung, oral cavity, larynx, oesophagus, stomach, pancreas, cervix, stomach, colon, rectum, ovary and bladder. Those in the population with the lowest intake of fruits and vegetables have almost double the normal incidence of these cancers. The data for cancers associated with hormone levels, however, showed less protection from fruits and vegetables. For breast cancer the protective effect was only about 30%. In the case of prostate cancer it is even less impressive.

Possible explanations of the protective effects manifested by high dietary intakes of fruit and vegetables (four to five helpings per day) are complex. However, compounds that can significantly modulate the activity of cellular enzyme systems involved in the detoxification, or activation, of potentially carcinogenic environmental chemicals are present in a number of plants. For example, allyl sulphide and diallyl sulphide, present in garlic and onions, can inhibit cytochrome P450 and activate glutathione transferase (see Chapter 5). Dithiothiones and isothiocyanates in many cruciferous vegetables such as broccoli, cabbage and cauliflower can protect experimental animals from carcinogens such as benzo[a]pyrene and dimethylbenzanthrene, probably through modulatory effects at the level of P450 enzymes. Other P450 modulators include quercitin (apples, onions) and ellagic acid (strawberries, raspberries, blackberries, walnuts, pecans). Fruits and vegetables also contain many substances capable of acting as antioxidants, such as tocopherols, carotenoids and flavonoids. At present there is an increasing interest in the protective role of dietary α-tocopherol (vitamin E), ascorbate (vitamin C) and β-carotene in lowering the incidence of a range of human tumours. The carotenoid, lycopene (from tomatoes), seems particularly protective against prostate cancer. In general, such plant antioxidants could diminish the opportunity for free radicals and reactive oxygen species derived from exogenous sources or generated intracellularly to exert their damaging effects on cellular DNA (see Chapter 5). Thus, despite the possibility that foodstuffs derived from fruit and vegetables contain natural and synthetic carcinogens (see also Chapters 2 and 11), the proportion of substances that exert beneficial effects appears to predominate.

8.6 *Markers of Cancer Risk from Environmental Carcinogens*

In order to minimize the risks of cancer, a reasonable strategy would be simply to steer clear of known carcinogens in the environment and workplace, take plenty of fruits and vegetables in the diet, reduce intake of red meat and animal fats, stay out of intense sunlight, and avoid tobacco smoke, excessive weight gain and high alcohol intake.

Another approach is to probe directly human tissue samples for specific biochemical clues, or 'markers', that would (a) reflect a person's past exposure to known environmental carcinogens, or (b) indicate the presence of damage to cell components that might lead to the development of cancers, or (c) point to a heightened sensitivity in particular persons to certain environmental carcinogens, or (d) reveal an inherited susceptibilty to the development of certain types of cancer.

For example, in terms of carcinogen exposure, it is possible to examine the DNA isolated from blood, lung or placenta samples for the presence of adducts with polycyclic aromatic hydrocarbons or aflatoxins. Oxidative DNA damage can be assessed from urine levels of thymine glycol or 8-oxo-guanosine. Damage specific to certain sites in the p53 gene can also be assessed in DNA from biopsy samples taken from breast, lung and liver. The pattern of mutation in that particular gene may identify the carcinogen responsible, and help to predict the risk of certain cancers. Blood samples can also provide DNA for the general assessment of inherited mutations in detoxification enzyme genes, antioxidant defence enzyme genes, protooncogenes and tumour suppressor genes. Mutations in the *BRCA1* gene, for example, may indicate inherited increased risk of breast and ovarian cancers. Absence of the *glutathione S-transferase M1* gene can indicate increased risk for lung and bladder cancer, and variations in the *cytochrome P4501A1* gene may be a useful indicator of elevated risk from lung cancer. Finally, blood samples can also be used for the determination of antioxidant levels (e.g. vitamins E and C, carotenoides, flavonoids). Low levels are markers of acquired susceptibility to certain cancers.

8.7 *Ageing*

Although environmental factors can be significant in the development of cancers as a result of their ability to damage crucial genes, this aspect may also be critical in the processes of ageing. Individuals are exposed to oxidants throughout their life, arising not only from environmental sources (Chapter 2) and endogenous sources such as mitochondria and inflammatory cells (Chapter 4). Many molecular gerontologists believe that this constant exposure to oxidants may be related to the basic processes of ageing. Although various antioxidant defences have evolved to combat this oxidant load, these are not perfect, and, as mammals age, their DNA accumulates oxidative damage. For instance an old rat (two years) has around two million oxidative lesions per cell. This is almost double the level detected in the DNA of rats a few weeks old. Mutations also accumulate with age, as does the rate of mutation. In humans the rate of somatic mutation in lymphocytes is some nine-fold higher in adults than in new-born infants. The ability to repair DNA damage also declines with age. This is particularly evident in cells from individuals with *Werner syndrome* which is a rare human disease, inherited as an autosomal recessive, with many symptoms of premature ageing. The defective gene in individuals with this syndrome has recently been cloned and its product appears to be a mutated DNA helicase, possibly involved in DNA repair.

When the level of oxidative damage in mitochondrial DNA in aged animals is compared with that suffered by nuclear DNA, there is a ten-fold difference. As mentioned in Chapter 6, there is speculation as to whether or not mitochondria can efficiently repair their DNA: if they cannot, this may account for the high level of damage to mitochondrial DNA observed. There may of course be other explanations. For example, mitochondrial DNA, unlike nuclear DNA of animal and plant cells, is not complexed with histones. On the other hand, the electron transport systems of mitochondria themselves are major sources of reactive oxygen species (see Chapter 4).

It has been argued that free radical damage to cell structures is a cause of ageing. Exogenous sources of free radicals such as ionizing radiation can reduce the lifespan of experimental animals and can cause changes similar to the ageing process. A positive correlation has been found between tissue concentrations of antioxidants such as vitamin E and carotenoids and lifespan in animals. Moreover, individuals who consume low levels of antioxidants in food have higher incidences of degenerative diseases of ageing such as cardiovascular disease, cancer and cataract. Reduction in the calorie content of the diet, aimed at diminishing mitochondrial activity, and thus superoxide generation, has been shown to increase the lifespan of experimental rats by over 40%. Dietary protein restriction also appears to extend lifespan,

and the age-related accumulation of free radical-damaged protein was slowed in rats receiving low-protein diets. The free radical hypothesis of ageing may be a means of explaining the cumulative changes resulting in increasing probability of death with age. It could be the price we paid for living in an atmosphere containing oxygen and using free radicals through the immune system as protection from infectious organisms.

A more specific view, however, is that age-related damage and mutation in mitochondrial DNA could be particularly critical. Age-associated changes in mitochondrial DNA including nucleotide mutations, oxidative base modifications and deletions have been observed. Such alterations could have a progressively deleterious effect on the ability of mitochondrial DNA to direct the construction of functional mitochondria. Indeed, a single mutation could initiate a recurring cycle of inhibited electron transport, leading to increases in superoxide formation and further mitochondrial DNA mutations. Defective function of the mitochondrial electron transport systems will subsequently lead to the depletion of cellular ATP supplies, and eventual cell and tissue dysfunction which could contribute to the symptoms of ageing.

Another potentially important concept is that of *longevity genes*. Mutant organisms that are especially long-lived may have alterations in genes specifying lifespan. Work with the nematode *Caenorhabditis elegans* has revealed an *age-1* mutant with a 70% increase in life expectancy and a doubling of lifespan. Further characterization of this mutant nematode revealed it to have elevated levels of the antioxidant defence enzymes superoxide dismutase and catalase, and to accumulate oxidative damage in its mitochondrial DNA at a lower level than in the wild type. Particularly relevant to environmental considerations was the increased resistance of the mutant to a variety of environmental stresses such as hydrogen peroxide, elevated temperature, ultraviolet radiation and the herbicide paraquat. The life span of fruit flies, *Drosophila melanogaster*, can be deliberately extended by genetically engineering the simultaneous overexpression of both Cu,Zn-superoxide dismutase and catalase. Such data and other studies suggest that lifespan is under some regulatory control. However, this control is not direct; rather it results from the passive outcome of alteration in the sensitivity of the organism to the environment, together with an increased ability to resist or repair environmental damage. Stress tolerance may well be the key to a longer life.

8.8 *Literature Sources and Additional Reading*

AGARWAL, M.L., TAYLOR, W.R., CHERNOV, M.V., CHERNOVA, O.B. and STARK, G.R., 1998, The p53 network, *Journal of Biological Chemistry*, 273, 1–4.

ALBERTINI, R.J. and O'NEILL, J.P., 1995, Monitoring for somatic mutations in humans, in Phillips, D.H. and Venitt, S. (Eds) *Environmental Mutagenesis*, pp. 341–66, Oxford: Bios Scientific Publishers.

AMES, B.N., GOLD, L.S. and WILLETT, W.C., 1995, The causes and prevention of cancer, *Proceedings of the National Academy of Sciences USA*, 92, 52258–65.

AMES, B.N., SHIGENAGA, M.K. and HAGEN, T.M., 1993, Oxidants, antioxidants and the degenerative diseases of aging, *Proceedings of the National Academy of Sciences USA*, 90, 7915–22.

AUTEXIER, C. and GREIDER, C.W., 1996, Telomerase and cancer: revisiting the telomere hypothesis, *Trends in Biochemical Sciences*, 21, 387–91.

BARINGA, M., 1998, Death by dozens of cuts, *Science*, 280, 32–34.

BRIDGES, B.A., 1997, Hypermutation under stress, *Nature*, 387, 557–58.

BURDON, R.H., GILL, V., BOYD, P.A. and RAHIM, R.A., 1996, Hydrogen peroxide and sequence specific DNA damage in human cells, *FEBS Letters*, 383, 150–54.

CAI, J. and JONES, D.P., 1998, Superoxide and apoptosis. Mitochondrial generation triggered by cytochrome c loss, *Journal of Biological Chemistry*, 237, 11401–14.

CARIELLO, N.F. and SKOPEK, T.R., 1993, *In vivo* mutation at the human HPRT locus, *Trends in Genetics*, 9, 322–26.

CARNEY, J.P., MASER, R.S., OLIVARES, H., DAVIS, E.M., LE BEAU, M., YATES III, J.R., HAYS, L., MORGAN, W.F. and PETRINI, J.H.J., 1998, The hMre11/hRad50 protein complex and Nijmegen breakage syndrome: linkage of double-strand break repair to the cellular DNA damage response, *Cell*, 93, 477–86.

CAVENEE, W.K. and WHITE, R.L., 1995, The genetic basis of cancer, *Scientific American*, 272, March, 50–57.

COLLER, H.A. and THILLY, W.G., 1994, Development and applications of mutational spectra technology, *Environmental Science and Technology*, 28, 478A–87A.

CROTEAU, D.L. and BOHR, V.A., 1997, Repair of oxidative damage to nuclear and mitochondrial DNA in mammalian cells, *Journal of Biological Chemistry*, 272, 25409–12.

CUTLER, R.G., 1991, Antioxidants and aging, *American Journal of Clinical Nutrition*, **53**, 373S–79S.

DARBY, S.C. and ROMAN, E., 1996, Links with childhood leukaemia, *Nature*, **382**, 303–304.

ELLEDGE, S.J., 1996, Cell cycle checkpoints: preventing an identity crisis, *Science*, **274**, 1664–72.

ENOCH, T. and NORBURY, C., 1995, Cellular responses to DNA damage: cell cycle checkpoints, apoptosis and the roles of p53 and ATM, *Trends in Genetics*, **20**, 426–30.

FORSTER, R., 1995, Measuring genetic events in transgenic animals, in Phillips, D.H. and Venitt, S. (Eds) *Environmental Mutagenesis*, pp. 291–314, Oxford: Bios Scientific Publishers.

GILCHREST, B.A. and BOHR, V.A., 1997, Aging processes, DNA damage and repair, *FASEB Journal*, **11**, 322–30.

GLICKMAN, B.W., KOTTURI, G., DE BOER, J. and KUSSER, W., 1995, Molecular mechanisms of mutagenesis and mutational spectra, in Phillips, D.H. and Venitt, S. (Eds) *Environmental Mutagenesis*, pp. 33–59, Oxford: Bios Scientific Publishers.

GROLLMAN, A.P. and MORIYA, M., 1993, Mutagenesis by 8-oxoguanine: the enemy within, *Trends in Genetics*, **9**, 246–49.

HARMAN, D., 1971, Free radical theory of aging: effect of the amount and degree of unsaturation of dietary fat on mortality rate, *Journal of Gerontology*, **26**, 451–57.

HENGARTNER, M.O., 1998, Death cycle and Swiss army knives, *Nature*, **391**, 441–42.

HOLLSTEIN, D., SIDEANSKY, D., VOGELSTEIN, B. and HARRIS, C.C., 1991, p53 mutations in human cancers, *Science*, **253**, 49–53.

HSU, I.C., METCALF, R.A., SUN, T., WELSH, J.A., WANG, N.J. and HARRIS, C.C., 1991, Mutational hotspot in the p53 gene in human hapatocellular carcinomas, *Nature*, **350**, 427–28.

JACKSON, S.P., 1996, The recognition of DNA damage, *Current Opinion in Genetics and Development*, **6**, 19–25.

JAZWINSKI, S.M., 1996, Longevity, genes and aging, *Science*, **273**, 54–58.

JONES, A.M. and DANGL, J.L., 1996, Logjam at the Styx: programmed cell death in plants, *Trends in Plant Science*, **1**, 114–19.

KANUGO, M.S. 1994, *Genes and Aging*, Cambridge: Cambridge University Press.

KING, R.J.B., 1996, *Cancer Biology*, Harlow: Addison Wesley Longman.

KORSMEYER, S.J., 1995, Regulators of cell death, *Trends in Genetics*, **11**, 101–105.

LANE, D., 1998, Awakening angels, *Nature*, 394, 616–17.

LEE, C.M., WEINDRUCH, R. and AITKEN, J.M., 1997, Age-associated alterations in the mitochondrial genome, *Free Radical Biology and Medicine*, 22, 1259–69.

LEFFELL, D.J. and BRASH, D.E., 1996, Sunlight and skin cancer, *Scientific American*, 275, July, 38–43.

LENGAUER, C., KINZLER, K.W. and VOGELSTEIN, B., 1998, Genetic instabilities in human cancers, *Nature*, 396, 643–9.

LI, R., YERGANAIN, G., DUESBERG, P., KRAEMER, A., WILLER, A., RAUSCH, C. and HEHLMANN, R., 1997, Aneuploidy correlates 100% with chemical transformation, *Proceeding of the National Academy of Sciences USA*, 94, 14566–71.

LITHGOW, G.J. and KIRKWOOD, T.B.L., 1996, Mechanisms and the evolution of aging, *Science*, 273, 80.

LOMBARD, D.B. and GUARENTE, L., 1996, Cloning the gene for Werner's syndrome: a disease with many symptoms of premature aging, *Trends in Genetics*, 12, 283–86.

MCBRIDE, T.J., PRESTON, B.D. and LOEB, L.A., 1991, Mutagenic spectrum resulting from DNA damage by oxygen radicals, *Biochemistry*, 30, 207–13.

MOXON, E.R. and THALER, D.S., 1997, The tinkerer's evolving tool-box, *Nature*, 387, 659–62.

NEEL, J.V., 1998, Genetic studies at the Atomic Bomb Casualty Commission–Radiation Effects Research Foundation: 1946–97, *Proceedings of the National Academy of Sciences USA*, 95, 5432–36.

PATLAK, M., 1997, Fingering carcinogens with genetic evidence, *Environmental Science and Technology*, 31, 190A–192A.

PERERA, F.P., 1996, Uncovering new clues to cancer risk, *Scientific American*, 274, May, 40–46.

PFEIFER, G.P., 1996, *Technologies for Detection of DNA Damage and Mutation*, New York: Plenum Press.

PHILLIPS, D.H. and FARMER, P.B., 1995, Protein and DNA adducts as biomarkers of exposure to environmental mutagens, in Phillips, D.H. and Venitt, S. (Eds) *Environmental Mutagenesis*, pp. 367–95, Oxford: Bios Scientific Publishers.

POLYAK, K., XIA, Y., ZWEIER, J.L., KINZLER, K.W. and VOGELSTEIN, B., 1997, A model for p53 induced apoptosis, *Nature*, 389, 300–305.

ORR, W.C. and SOHAL, R.S., 1994, Extension of life span by overexpression of superoxide dismutase and catalase in *Drosophila melanogaster*, *Science*, 263, 1128–30.

OSAMI, S.A. and YE, X.S., 1997, Targets of checkpoints controlling mitosis: lessons from lower eukaryotes, *Trends in Cell Biology*, 7, 283–88.

SCHULL, W.J., 1998, The somatic effects of exposure to atomic radiation: the Japanese.experience, 1947–1997, *Proceeding of the National Academy of Sciences USA*, 95, 5437–41.

SHIGENAGA, M.K., HAGEN, T.M. and AMES, B.N., 1994, Oxidative damage and mitochondrial decay in aging, *Proceeding of the National Academy of Sciences USA*, 91, 10771–78.

SOHAL, R.S. and WEINDRIUCH, R., 1996, Oxidative stress, caloric restriction and aging, *Science*, 273, 59–63.

TRICHOPOULOS, D., LI, F.P. and HUNTER, D.L., 1996, What causes cancer? *Scientific American*, 275, September, 50–57.

WALDRON, K.W., JOHNSON, I.T. and FENWICK, G.R., 1993, *Food and Cancer: Chemical and Biological Aspects*, Cambridge: The Royal Society of Chemistry.

WALLACE, D.C., 1997, Mitochondrial DNA in aging and disease, *Scientific American*, 277, 22–29.

WEBB, D.K., EVANS, M.K. and BOHR, V.A., 1996, DNA repair fine structure in Werner's syndrome cell lines, *Experimental Cell Research*, 224, 272–78.

WEINERT, T., 1998, DNA damage and checkpoint pathways: molecular anatomy and interactions with repair, *Cell*, 95, 555–58.

WILSON, D.M. and THOMPSON, L.H., 1997, Life without DNA repair, *Proceedings of the National Academy USA*, 94, 12754–57.

WIMAN, K.G., 1997, p53: emergency brake and target for cancer therapy, *Experimental Cell Research*, 237, 14–18.

WISEMAN, H. and HALLIWELL, B., 1996, Damage to DNA by reactive oxygen and nitrogen species: role in inflammatory diseases and progression to cancer, *Biochemical Journal*, 313, 17–29.

WYLLIE, A., 1997, Clues in the p53 murder mystery, *Nature*, 389, 237–38.

YU, B.P., 1996, Aging and oxidative stress: modulation by dietary restriction, *Free Radical Biology and Medicine*, 21, 651–68.

ZEIGER, E., 1995, Mutagenicity tests in bacteria as indicators of carcinogenic potential in mammals, in Phillips, D.H. and Venitt, S. (Eds) *Environmental Mutagenesis*, pp. 107–19, Oxford: Bios Scientific Publishers.

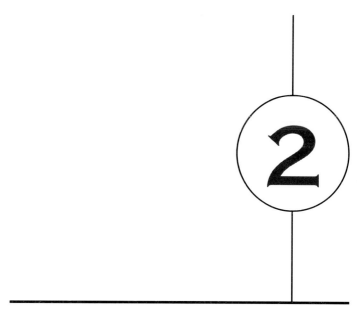

ENVIRONMENTAL STRESS:

GENOMIC RESPONSES

CHAPTER **9**

Gene Activation and Environmental Stress

KEY POINTS

- Organisms harbour a variety of stress genes whose expression can be elevated following exposure to a range of environmental threats.

- This leads to the production of special cellular stress proteins that play important roles in helping organisms to resist particular environmental threats.

- The products of such stress genes include heat-shock proteins (that confer thermal resistance); antifreeze proteins; osmoregulators; UV-protectors, metallothioneins and toxic metal metabolizing enzymes; transporter ATP-ases; pathogen defence systems; antioxidant defence enzymes; detoxifying enzymes and other specialized enzymes of cellular metabolism.

- Certain chemicals in the environment can mimic, or disrupt the biological effects of, natural hormones. This different type of stress can result in the inappropriate regulation of genes required for key processes in normal animal development, with potentially serious biological consequences.

Since their earliest appearance on earth, living organisms have been exposed to environments which altered gradually and sometimes rapidly. Long-term adaptation to environmental change has been possible through processes of mutation and natural selection. Another more immediate means of protection against environmental adversity involves the induction of stress gene expression as a pleiotropic response to stress. The products of stress genes play key roles in cellular homeostasis through protective and adaptive functions, enabling organisms to survive different types of environmental onslaught. These stress responses are orchestrated at the gene level and characteristically can be induced both quickly and repeatedly. They are thus particularly valuable to organisms living under rapidly altering environments. Moreover, a large proportion of these responses are seen in all organisms from bacteria to humans and are likely to have originated soon after primitive cells appeared on earth. With the dramatic advances in molecular biological techniques in recent years, it has been possible to conduct detailed investigations of the underlying inductive mechanisms. Although the stress of DNA damage elicits SOS responses in bacteria and the activation of cell division control genes by p53 in eukaryotes (see Chapter 8), there are many environmental threats that have been shown to bring about the activation of other specific stress genes. These include temperature shifts, oxidative stress, ultraviolet radiation, water and salt stress, toxic metals, anaerobiosis and pathogenicity.

9.1 *High-temperature Stress*

Temperatures are continually changing within the environment, both during the day and with the season. These fluctuations can have considerable impact on many different biochemical and physiological processes such as respiration and photosynthesis, as well as other cellular protein and membrane functions. In 1962 it was found that when the temperature at which fruit fly larvae were developing was raised from the normal 25°C to 32°C, a number of new sites on the giant chromosomes of the larval cells became activated. This curious finding suggested that increased temperature induces the expression of specific genes. This was confirmed some years later with the characterization of several novel proteins that were present in the larvae following the temperature rise. This phenomenon is not peculiar to fruit flies; it can be initiated in all types of organism from bacteria to humans, and is referred to as the *heat-shock response*. Subsequent investigations have shown that the novel proteins identified in the heat-shocked cells can also be detected in normal cells, at very much lower concentrations. Nevertheless, they are still collectively referred to as the *heat-shock proteins (Hsps)*.

9.1.1 HEAT SHOCK PROTEINS

Because the functions of individual Hsps were initially poorly understood, it has been conventional to refer to individual proteins according to their approximate molecular weights. For example, members of the Hsp70 class have molecular weights in the region of 70 to 73 kDa. Most of the Hsps are relatively acidic polypeptides with isoelectric points between pH5 and 7. Table 9.1 provides a list of the major classes of Hsp found in mammalian cells and gives an indication of their predominant intracellular location. Increased Hsp synthesis in mammalian cells can be observed following, for example, continuous exposure to temperatures between 40°C and

TABLE 9.1 *Heat-inducible proteins of mammalian cells*

Class of Hsp	Location	Comment	Bacterial homologue
Hsp110	Nucleolus	Normal cell component	
Hsp90	Cytoplasm	Normal cell component, associates with steroid receptors	LtpG
Hsp70	Cytosol nucleus Nucleolus Mitochondria Endoplasmic reticulum	Hsp73 – normal cell component Hsp72 – normal cell component but highly heat-inducible	DnaK
Hsp60	Mitochondria	Normal mitochondrial protein	GroEL
Hsp47	Membrane glycoprotein	Affinity for collagen	
Hsp32	Cytoplasm	Haem oxygenase	
Hsp28	Golgi complex	Normal cell component – phosphorylation status depends on growth stimuli	
Hsp10	Mitochondria	Normal mitochondrial protein	GroES

42°C. Above such temperatures all synthesis of protein in mammalian cells declines significantly.

In plants there is a threshold temperature above which the synthesis of Hsps is elevated. Such threshold temperatures vary with species rather than with environment. For instance, Hsp induction occurs in maize at 30°C, whereas in millet induction requires temperatures of 40°C and above, although all protein synthesis declines markedly beyond 50°C. The heat-shock proteins of plants are represented by the Hsp 90, Hsp70 and Hsp60 classes. In plants there is also a wide range of low molecular weight Hsps between 14 and 30 kDa. These are more predominant in dicotyledons than in monocotyledons, and are mainly found in the cytoplasm as components of heat-shock granules. Members of the Hsp60 class are found not only in mitochondria but also in chloroplasts. Plant Hsp70s are also found in chloroplasts, as well as in mitochondria, endoplasmic reticulum, cytoplasm and nuclei. In evolutionary terms the Hsp70s are highly conserved, the human Hsp70 being 73%

homologous to the fruit fly Hsp70 and 68% homologous with maize Hsp70.

The response in bacteria to an increase in growth temperature resembles that encountered in plants and animals. Almost immediately after elevation of growth temperature in *E. coli*, there is an accelerated synthesis of around 17 Hsps. For example, when cells are shifted from 30°C to 40°C the rates of their synthesis increases between five- and 20-fold. Three of these proteins, DnaK, DnaJ and GrpE, which are required for growth at high temperatures, also play a vital role in the replication of infecting bacteriophage through protein–protein interactions and ATP hydrolysis. One of these, DnaK, is now recognized as the bacterial equivalent of mammalian and plant Hsp70. Two other heat-induced bacterial proteins, GroES and GroEL, are also required for growth at elevated temperatures as well as for bacteriophage replication. GroEL shows homology with mammalian Hsp60; GroES appears to be equivalent to mammalian Hsp10.

Studies have also been carried out in archaeobacteria that thrive under extreme environmental conditions which preclude the survival of other life forms. They probably evolved under environmental conditions where temperatures may have exceeded 100°C, and some species currently inhabit similar niches. It could be that possession of heat-shock genes provided archaeobacteria with a selective advantage and played a role in their evolution. In the case of the extreme thermophiles, *Sulpholobus acidocaldarius* and *Sulphobus shibatae*, raising the temperature from normal growth temperature of 70°C to 85°C causes the decline in synthesis of many proteins but induces the synthesis of a limited number (3–5) of Hsps, although no 70 kDa equivalents of DnaK protein were observed in these species. In *Pyrodictium occultum* cells, which normally grow at 102°C, increasing the temperature to 108°C leads to an increase in two heat shock polypeptides, but again no increment in any 70 kDa proteins. In general it is evident that the genes of extreme thermophiles are able to function at very high temperatures. Moreover, the changes in patterns of proteins expressed in their response to heat shock is concomitant with the establishment of thermotolerance.

9.1.2 THERMOTOLERANCE AND THE HEAT-SHOCK PROTEINS

Human and other mammalian cells die after exposure to temperatures above 43°C. Heat has toxic effects towards membrane function, metabolism, cytoskeleton relationships and ionic homeostatsis, as well as cell cycle progression and macromolecular synthesis. If the exposure is sub-lethal (i.e. temperatures below 43°C), a notable feature is the rapid, but transient, reprogramming of cellular activity to ensure survival during the period of stress, to protect essential cell components against heat damage and to permit a rapid resumption of normal cellular activities during the recovery period. Such exposure to near lethal temperatures usually leads to a degree of adaptation so that a previously lethal hyperthermic exposure is tolerated. This is the phenomenon of *thermotolerance*, and heat treatments that induce cellular thermotolerance in animal cells also induce the intracellular accumulation of heat-shock proteins.

In plants the effect of high temperature is primarily on photosynthetic functions. The heat tolerance of higher plants coincides with the thermal sensitivity of primary photochemical reactions occurring in the thylakoid membrane systems. Most plants are stable within a range of 10°C to 35°C, but significantly higher tolerance limits can be found among plants adapted to warm climates. Heat tolerance limits are determined by the thermal stability of structural entities used for physiological processes. Mitochondria appear to be more stable than thylakoids, and respiration is known to increase with temperature beyond a level where photosynthesis declines. There is a large variability in heat tolerance limits among different plants. Adaptation limits can be the result of natural selection of genotypes (genotypic acclimation). Each genotype can also adjust its tolerance limit to seasonal changes in thermal environment (phenotypic adaptation), although the ability to achieve this varies with different plants. Long-term acclimation is superimposed by another type of adaptation occurring in a time range of a few hours. This may serve to adjust leaves to peak temperature during a day. The tolerance limit can increase by several degrees with a half-life of a few hours. This short-term adaptation requires protein synthesis and appears to be related to the accumulation of Hsps in plant structures including chloroplasts.

There are now many data to indicate a strong kinetic correlation between the induction of Hsps and the development of thermotolerance. Moreover, heat-resistant mammalian cell lines usually express elevated levels of various Hsps. Experiments involving microinjection of antibodies to Hsps into mammalian cells, or transfection of plasmids aimed at bringing about high levels of *hsp* gene expression, also support the view that Hsps reduce the sensitivity of cells to the effects of high-temperature stress.

Although many of the functions of Hsps are still the subject of intense investigation, it is now clear that an important role of Hsps is to prevent improper folding or aggregation of cellular proteins, or their subunits. It has been suggested that Hsps reversibly interact with internal regions of pre-existing cellular proteins that are exposed as a result of thermal denaturation during exposure to heat stress. This interaction results in the facilitation of correct folding of the aberrant proteins and possibly the reassembly of multiprotein structures damaged by heat.

9.1.3 HEAT-SHOCK PROTEINS AS MOLECULAR CHAPERONES

Within a normal cell it is crucial that polypeptide chains are correctly folded, processed, localized, and, in certain cases, complexed with other polypeptides, in order to carry out their biological function properly. As newly synthesized polypeptide chains emerge from ribosomes they are very likely to make premature contact with other protein domains because of the naturally high cytosolic protein concentration. Such premature interaction among protein domains must be avoided to prevent misfolding. Additionally, it has become clear that proteins that have to cross biological membranes must be unfolded. As a result, there is again the opportunity for inappropriate interactions between protein domains. Such constraints have suggested the need within cells for some type of 'chaperoning' activity to

keep proteins in an unfolded state and at the same time block undesirable interactions. Recent interest has focused on a set of cellular proteins referred to as *chaperones*. Their cellular role is to maintain and protect the unfolded state of newly synthesized proteins. They prevent them from misfolding or aggregating, then permit them to cross biological membranes and allow them to fold properly, thus leading to correct oligomerization. Critically, molecular chaperones act only to provide an environment in which a polypeptide can fold. They do not convey any information essential for the process; rather they assist in the process of polypeptide self-assembly. Since the same chaperones are able to assist in the folding of a wide variety of polypeptides, it is assumed that unfolded or partially folded intermediates share common structural features which allow them to associate with the same chaperone. Once a protein has achieved its final conformation, these structural features are likely to be buried within the interior of the protein, thus rendering the protein no longer capable of interaction with a chaperone.

Particularly relevant to the issue of heat stress is the fact that certain of the cellular chaperones were originally discovered as Hsps. Under the particular conditions of heat stress these molecular chaperones protect other proteins from heat denaturation, or, if damage has occurred, disaggregate them and allow them to refold back to an active form. In summary, molecular chaperones, *which include Hsps as an important subset*, appear to have evolved as part of a cellular safety or rescue system.

The family of *Hsp70 proteins* have a relatively high abundance in a wide variety of cellular compartments, and exhibit an extensive promiscuity in binding to unfolded polypeptides. As a group, the Hsp70 proteins are universally conserved, members being at least 50%

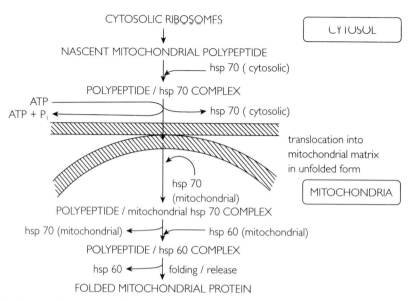

FIGURE 9.1 *Involvement of Hsps in translocation and folding of mitochondrial proteins*

identical to each other in terms of amino acid sequence. In *E. coli* the family is represented by the DnaK protein. In mammalian cells there are: Hsp73, which is expressed constitutively and present in the cytosol and nucleus; the heat-inducible Hsp72, present in the cytosol, nucleus, nucleolus; and Hsp70s in mitochondria and endoplasmic reticulum. A weak ATPase activity is associated with the N-terminal part of Hsp70 molecules, and it appears that a binding site accommodates the polypeptide ligand. There is a perfect correlation between ATP hydrolysis and the release of polypeptides. Hsp70s also exibit an ability to protect various enzymes from thermal inactivation and to disaggregate protein aggregates. In general, the interaction of a particular Hsp70 with a target protein, or proteins, is transient, with the release of the target protein being dependent on ATP hydrolysis. Importantly, in cells experiencing stress the interaction is no longer transient.

Hsp70s interact transiently and post-translationally with nascent polypeptides to facilitate their proper folding. They are also essential in facilitating the translocation of proteins across biological membranes such as the endoplasmic reticulum and those of mitochondria (Figure 9.1). In the case of mitochondria, some 95% of the proteins are made as precursor proteins on the ribosomes of the cytosol and are then imported into the four mitochondrial subcompartments (outer membrane, intermembrane space, inner membrane and matrix). While cytosolic Hsps70s play a key role in maintaining the precursor proteins in an unfolded transport-competent conformation, the Hsp70 of mitochondria is involved in the translocation of precursor proteins through contact sites between mitochondrial inner and outer membranes. The ultimate folding of the imported mitochondrial proteins then involves the participation of the Hsp60 of the mitochondrial matrix (see Figure 9.1).

Members of the *Hsp60 family* have been well conserved throughout evolution and are found in bacteria as well as in the mitochondria and chloroplasts of plants and animals. In *E. coli* this family is represented by the products of the *groEL* gene. The amino acid sequence of the GroEL protein is around 50% identical with that of the Hsp60 of eukaryotes and the ribulose–bisphosphate carboxylase/oxygenase-binding protein of plant chloroplasts.

Both GroEL and eukaryotic Hsp60 can bind unfolded polypeptides and progressively assist their proper folding. GroEL is particularly well studied and functions as a large molecular complex of 14 identical polypeptide subunits arranged in two stacked rings resembling a double doughnut. Each ring comprises seven subunits forming a chamber surrounding a central cavity. At present it is thought that polypeptides in intermediate stages in the folding process become sequestered within the central cavity through the interaction of certain of their hydrophobic amino acid residues with hydrophobic patches in GroEL subunits. There they are prevented from undergoing misfolding, or aggregating with other proteins. Once in the central cavity, the partially folded protein is believed to cycle alternately between states where it is bound to the walls of the chamber and when it is free. These changes in the binding properties of GroEL are believed to be driven by the hydrolysis of ATP. Each time it is released from the chamber walls, the polypeptide appears to take a step towards its final conformation. Once the folding steps have been completed and the hydrophobic binding regions have become internalized, the polypeptide is no longer able to bind to the chamber wall and is subsequently released from the GroEL complex. Additional studies have indicated that another bacterial protein, GroES, may bind to the top and bottom of the GroEL complex and close off the inner chamber during the folding process. The eukaryotic homologue of GroES is thought to be Hsp10 (see Table 9.1).

Hsp90 proteins appear to associate with a number of key of proteins within cells. These include members of the steroid hormone receptor family, the oncogene product pp60[src] and a kinase that phosphorylates the eukaryotic protein synthesis initiatioin factor, EF2. Because of this, Hsp60s may play important regulatory roles in terms of cell function. In bacteria the family is represented by the LtpG protein, which is 40% identical with its eukaryotic cell counterparts.

In general it is likely that Hsp90s play a chaperoning role similar to Hsp60s and Hsp70s. Thus, they may bind to the nascent polypeptides mentioned above and either escort them to their proper cellular location or maintain them in a form for correct folding. In the particular case of the steroid hormone receptors, the complex with Hsp90 also includes Hsp70. In the absence of the steroid hormone, the receptor is kept inactive as part of this complex. Upon steroid addition, the Hsp90 is released and the remaining Hsp70–steroid receptor complex dissociates upon ATP hydrolysis, allowing the formation of a functional steroid receptor dimer. Overall, it appears that the initial capture and inactivation of the steroid receptor is an obligatory prelude to its correct folding and assembly.

9.1.4 HEAT-SHOCK GENE ACTIVATION

While increases in the rates of Hsp gene expression are generally in proportion to the severity of the heat stress, the inducibilty of individual Hsps can vary widely. The regulation of Hsp genes is controlled by heat shock transcription factors (HSFs). These have the ability to recognize modular sequences within the promoter regions upstream of Hsp genes. These are referred to as heat-shock elements (HSEs) and contain three or more five-base-pair units having the general consensus sequence NGAAN (Figure 9.2).

......**GAA**CCCCTG**GAA**TATTCCC**GAC**......   proximal HSE
......CTTG**GGG**ACCTTAT**AAG**GGCTG.....

.......**GTG**AATCCCA**GAA**GACTCTG**GAG**AGTTC.....   distal HSE
........CACTT**AGG**GTCTTCT**GAG**ACCTCTC**AAG**.....

FIGURE 9.2 *The nucleotide sequence of two HSEs located respectively around 100 (proximal) and 180 (distal) base-pairs upstream of the transcription start site of a human hsp70 gene (see also Figure 9.3): the nucleotides in bold type represent components of the five base-pair consensus sequences*

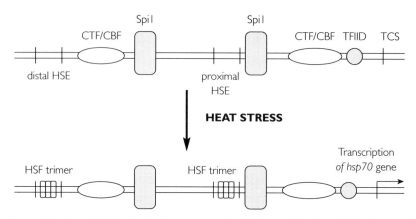

FIGURE 9.3 *Transcription factor binding sites in the promoter region of a human hsp70 gene. TFIID is normally bound to the TATA box sequence at −30 base pairs from the transcription site TCS, Spi1 to GGCGGG sequence elements, and CTF/CBF to CCAAT box sequences. However, after heat stress trimeric HSFs also bind to the promoter at both the proximal and distal HSEs (at −100 and −180 base-pairs respectively from the TCS)*

All HSFs have a conserved DNA-binding domain near the N-terminus and, importantly, they are present in both stressed and unstressed cells, their rate of expression being unaffected by heat stress. In unstressed cells, HSF is in an inactive form and stress is required to achieve the conversion to a transcriptionally competent form. Following heat stress, HSF, in eukaryotes other than yeasts, forms homotrimers which are necessary to facilitate HSE translocation to the nucleus and binding to HSE sequences upstream of the transcription start sites of *hsp* genes. In yeasts HSF is already trimeric and capable of DNA binding even in the absence of stress. Each subunit of an HSF trimer is believed to bind to a single NGGAN sequence of an HSE, and binding is highly cooperative. Nevertheless, it is clear that the formation of trimers is insufficient to make HSF capable of stimulation *hsp* gene transcription. A further event is yet required, but at present the nature of this additional event is not clear, although phosphorylation of HSFs may be relevant.

In normal unstressed human cells a number of transcription factors are bound to various sequence elements in the DNA of the upstream promoter region of a human *hsp70* gene (Figure 9.3). For example, the transcription factor TFIID is bound to the 'TATA box' element located at −30 base-pairs from the transcription start site (TCS). Other transcription factors bound are Spi1 at GGCGGG sequences and CTF/CBF at 'CCAAT boxes'. Following heat stress, such as exposure of the human cells to 42°C, there is the additional binding of trimeric HSFs to the proximal (−100) and distal (−180) HSEs. Importantly,

TABLE 9.2 *Some inducers of heat-shock protein gene activation*

INDUCER	COMMENT
Group 1	*Cause synthesis of inactive proteins*
Amino acid analogues	Yield proteins with improper amino acid side-chains
Puromycin	Results in truncated proteins due to premature termination of protein synthesis
Ethanol	Aberrant proteins produced due to errors in translation
Group 2	*Cause destruction of conformation of native proteins*
Heat treatment	By increasing the rate of protein unfolding
Heavy metal ions	By directly, or indirectly, causing protein thiolation
Arsenite	or protein–protein disulphide formation, through
Thiol reactive compounds	redox state alterations
(e.g. p-mercuribenzoate, iodoacetate, diamide)	By acting to change the redox state of cells and
Recovery from anoxia	hence bring about covalent modification of proteins
Hydrogen peroxide	By causing changes in ATP levels and the cellular
Superoxide	redox state
Amytal	
Antimycin	
Azide	
Dinitrophenol	
Rotenone	

in addition to heat stress, an extremely diverse range of agents and treatments can also activate the transcription of *hsp* genes. Some of these are presented in Table 9.2.

The two groups of inducers listed in Table 9.2 can bring about the formation of *non-native proteins* by different mechanisms. This has suggested that *hsp* gene activation may be triggered by the accumulation of non-native proteins, and has led to the view that Hsp70 itself might act as a sensor of such an accumulation.

9.1.5 HSP70 AS A SENSOR OF NON-NATIVE PROTEINS AND REGULATOR OF *HSP* GENES

The demonstration that accumulation of non-native proteins activates the induction of *hsp* genes provides a conceptual framework for the view that Hsp70 proteins may be an important regulator of these genes. A basic requirement of a cellular signalling system is a molecule capable of sensing abnormally elevated levels of non-native proteins and conveying this information to the HSFs. Hsp70 proteins are ideal candidates. They are a family of proteins whose main function is directly related to their ability to recognize non-native conformations in proteins, and a simple regulatory mechanism has been proposed. In this, the formation of HSF trimers is prevented in non-stressed cells by the formation of HSF–Hsp70 heterodimers (or higher order complexes containing HSF). When cells are stressed they accumulate abnormal amounts of non-native proteins. These then compete with HSF for binding of Hsp70. In

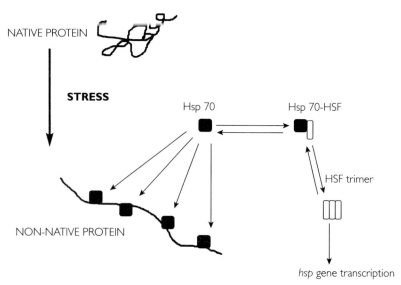

NATIVE PROTEIN

STRESS

Hsp 70

Hsp 70-HSF

HSF trimer

NON-NATIVE PROTEIN

hsp gene transcription

FIGURE 9.4 *A model illustrating the possible role of Hsp 70 in hsp gene regulation*

the face of this competition, the proportion of unbound HSF, and thus the possibility of its trimerization, increases (Figure 9.4). Such trimerized HSF then participates in the stimulation of *hsp* gene transcription. If the conditions of stress are removed, Hsps continue to be made and non-native proteins are either restored to their native conformation or destroyed by cellular proteolytic systems. In due course, the cellular levels of Hsp70s available to bind HSF increase to normal prestress concentrations, which enables HSF–Hsp70 heterodimers to reform.

9.1.6 PHYSIOLOGICAL INDUCERS OF *HSP* GENES

Despite the attractiveness of the proposal involving Hsp70 as sensor of proteotoxic stress and its role in signalling *hsp* gene transcription through interactions with HSFs, it should be emphasized that there can be significant basal levels of *hsp* gene expression in various cells that are unstressed. Hsps are also induced in animal cells by physiological

non-stress situations. For example, growth-regulated expression of *hsp70* genes is observed in previously quiescent cells following serum or cytokine stimulation. Viral infection can also induce increased *hsp* gene expression. It appears that specific sequence elements in the proximal region of the *hsp70* gene promoter (from −1 to −68 base pairs) are necessary for the maintenance of basal 'constitutive' levels of expression and the physiological response to growth signals and viral infection.

9.1.7 COMPARISON OF EUKARYOTES AND PROKARYOTES

Although the response to heat stress is universal, its quality and magnitude vary widely in nature. Eukaryotes often possess two copies of most *hsp* genes, one regulated by heat stress induction and one expressed continuously. Prokaryotes, in contrast, possess only a single copy of each heat-shock gene. Due to the essential nature of most of the prokaryotic heat-shock genes, this necessitates their continuous expression under most conditions. In *E. coli* the major heat-shock genes are under the control of the *htpR(rpoH)* gene, which encodes a novel sigma factor, σ^{32}. Each *E. coli* heat-shock gene is transcribed from a promoter which is recognized by the bacterial RNA polymerase containing σ^{32}. Transcription of *htpR* itself is under temperature regulation, and mutations in the *dnaK*, *dnaJ* or *grpE* genes result in 'overexpression' of the heat-shock response, demonstrating that the heat-shock proteins from these genes can negatively regulate the transcriptional response in bacteria.

9.2 *Low-temperature Stress*

Numerous physical environments place severe water restrictions on organisms. A particular example arises from the phase transition from liquid water to solid ice. With the exception of mammals, most organisms have little control over their body temperatures and are vulnerable to the hazard of crystallization of water into ice within their tissues. Ice causes physical injuries to cells due to forces of shearing, puncture or expansion. Ice that penetrates cells also wreaks havoc with subcellular compartmentation and microarchitecture.

Initially a freezing front sweeps through the extracellular spaces of an organism without penetrating cells. This can cause extensive damage due to the extreme hyperosmotic stress it imposes. Ice is a crystal of pure water and excludes from its lattice all the dissolved solutes. This elevates their concentration in the extracellular fluids that remain, and stimulates the osmotic outflow of water from cells. As a result cells shrink to a fraction of their original volume. Many organisms that cannot avoid exposure to sub-zero temperatures, for example by habitat selection or sheltered hibernation sites, instead employ strategies

to suppress the temperature at which water freezes. Perhaps the most sophisticated of these are the manufacture of *antifreeze proteins* and the synthesis of low-molecular weight *colligative protectants* such as polyols to increase the osmolality of extracellular fluids.

9.2.1 ANTIFREEZE PROTEINS

Antifreeze proteins (AFPs) have arisen independently in several phylogenetically distinct groups and have been most studied in cold water marine fishes and cold-tolerant insects. Four distinct classes of fish AFP are known, with other types present in insects, although they are structurally different, the mechanisms whereby they lead to ice growth inhibition are similar. AFPs are glycoproteins with the ability to absorb directly onto the ice crystal lattice through a hydrogen bonding network to the surface water molecules, involving the polar side-chains of amino acids and hydrophilic carbohydrate groups. While not preventing ice formation, they break up the crystal growth plane and inhibit growth by adhering to the prism face of the crystals. By this means they bring about a non-colligative reduction in freezing points without changing melting points. In the case of fish APFs, the reduction in freezing point that can be brought about in body fluids is 1–1.5°C. AFPs from terrestrial insects are somewhat more powerful, and can lower the freezing points of haemolymph by 6–9°C.

9.2.2 COLLIGATIVE PROTECTANTS

Colligative protectants are often used by organisms where more powerful protection is required from freezing conditions. Although some insects use ethylene glycol, the polyol glycerol is the most commonly used antifreeze, often accounting for 20% of the fresh weight of insects in the winter. Glycerol can be synthesized from reserves of glygogen laid down during feeding in late summer. Manufacture is induced by the activation of glygogen phosphorylase at low temperatures (around 5°C), with most carbon directed through the hexose monophosphate shunt to yield glyceraldehyde and NADPH, the substrates for the glycerol dehydrogenase reaction. In combination with the presence of AFPs, glycerol can allow many insect species to maintain their body fluids in a liquid state down to temperatures approaching −40°C. Other polyols, such as threitol, erythritol and sorbitol, as well as some monosaccharides and disaccharides, are used as antifreezes by some species.

9.2.3 COLD ACCLIMATION AND LOW-TEMPERATURE RESPONSIVE GENES

While low temperature will elevate the activity of glygogen phosphorylase and glycerol production in insects, evidence of a *cold-shock response* has been observed in bacteria. In both *E. coli* and *B. subtilis*, an abrupt down-shift in growth temperature increased the production of major *cold-shock proteins*, CS7.4 and CspB respectively. CS7.4 is a transcriptional activator of at least two genes and, importantly, both CS7.4 and CspB share considerable sequence homology with a nucleic acid binding domain of a class of eukaryotic gene regulatory factors (Y-box factors). However, the role of these particular regulatory cold-shock proteins in the context of bacterial stress physiology is not yet fully understood.

In the case of plants, mechanisms have evolved that allow acclimation to cold temperature in certain cases. For instance, in many temperate plants, a period of exposure to a low positive temperature will acclimate the plants to withstand a subsequent freezing stress. The process is complex, and cellular and metabolic changes that occur during cold acclimation include increased levels of polyol sugars, soluble proteins (some with cryoprotective activity), as well as the appearance of new isoforms of proteins and alteration of lipid membrane composition. Many of these changes are regulated by changes in expression of *low-temperature responsive (LTR) genes*. The ability to acclimate to frost tolerance is under genotypic control, and low temperature (6–2°C) will cold-acclimate plants that are genetically competent (e.g. winter cereals), but will damage chilling-sensitive plants (e.g. maize). However, not all chilling-tolerant plants are able to acclimate to frost tolerance (e.g. spring cultivars). Besides acclimation for frost tolerance, a prior period of exposure to low positive temperature may also acclimate chill-tolerant plants for growth at low temperatures.

A primary site of injury during freezing is the plasma membrane, and membrane lipid composition is believed to be significant in tolerance to low temperatures. Plants grown at low temperatures manifest an increase in lipid unsaturation, but changes in lipid composition and the ratio of lipids to protein may also alter. In plants that can normally acclimate to frost tolerance, mutations that cause deficiencies in the production of unsaturated fatty acids result in plants that are chill-sensitive and can no longer acclimate to frost tolerance. In experiments where levels of unsaturated lipids were manipulated by genetic transformation, the chill-sensitivity of the resulting plants was altered. Transgenic plants with genes that cause increased levels of unsaturated fatty acids (e.g. ω-3 fatty acid desaturase) exhibit an increase in resistance to chilling temperatures. Overall there is now considerable evidence for the involvement of polyunsaturated lipids in the resistance of membranes to low-temperature responses in plants, although this may not be directly involved in frost acclimation.

A number of plant genes are known to be activated specifically by acclimation treatments. Significantly, many of these LTR genes are also activated by droughting (see below). Since, during freezing, ice crystals form outside cells, freezing and droughting can both be viewed as dehydration stresses. Relatively little is known of the the function of LTR gene products. Some appear to be low-temperature tolerant isoforms of vital 'housekeeping' proteins. Other protein products are thought to play a cryoprotective role, possibly with properties similar to the antifreeze proteins of insects and fish. Surprisingly, none of the temperate plant LTR genes so far studied appears to be involved directly in lipid synthesis, although one LTR gene family from barley that may encode a group of non-specific lipid transfer proteins has been identified. These may be involved in wax synthesis or secretion. On the other hand, in the cyanobacterium *Synechocystis* sp. PCC6803, a downward shift in temperature of 10°C to 15°C causes the cessation of fatty acid synthesis but a rise in their level of desaturation. Once the degree of unsaturation reaches a certain level, the cyanobacteria begin to grow again and synthesize fatty acids. The increase in desaturation corresponds to an up-regulation in the expression of genes for fatty acid desaturases.

Besides their activation by low temperatures (or drought, as described below), a significant number of temperate plant LTR genes are also up-regulated by *abscisic acid (ABA)*. This plant growth regulator is also known to induce freezing tolerance in certain plants. While the upstream promoter region of a number of LTR genes so far studied contains a *low-temperature responsive element (LTRE)* with the consensus sequence *A/GCCGAC*, several

of the promoters also contain *G-box elements, CACGTG*, which correspond to the core nucleotide sequence motif of *ABA-responsive elements (ABREs)*. A possible mechanism involving signal transduction emanating from the plasma membrane ABA receptor implicates increased calcium levels in the cytosol. In turn this may modulate protein phosphatase activity, and thus the phosphorylation status of key regulator proteins relevant to the activation of LTR genes responsive to ABA that are involved in the cold acclimation responses (Figure 9.5).

Although low-temperature exposure also increases plant cell cytosolic calcium levels, an important question is the means whereby low temperature itself can trigger the signalling pathways leading to the activation of LTR/ABA responsive genes. There have been speculations that the sensor for perception of low temperature may be located in the plasma membrane. A physical phase transition may occur in the microdomains of the plasma membrane upon a downward shift in temperature, and that possible sensor molecule, perhaps akin to the histidine kinase that senses changes in osmolarity in yeasts, detects a change from the liquid-crystalline to the gel state. In turn this sensor molecule may undergo a conformational change, or a cycle of phosphorylation, as the primary event in the transduction of the low-temperature signal.

In addition to the ability to up-regulate LTR genes in the face of chilling adversity, many plant groups in colder climates manifest increased degrees of *ploidy*. It has been speculated that this may have led to comparatively richer gene pools enabling certain plants to adapt more readily to novel habitats in such climatic conditions.

In summary, while considerable information is accumulating regarding a wide range of LTR genes and their promoters, as yet comparatively little is known regarding the function

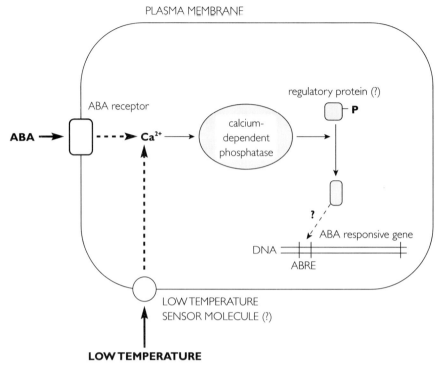

PLASMA MEMBRANE

FIGURE 9.5 *A highly schematic model indicating the possible participation of abscisic acid (ABA) in the acclimation of plants to low temperatures. Interaction of ABA with its plasma membrane receptor triggers the elevation of intracellular calcium concentration. In turn this modulates the activity of a calcium-dependent protein phosphatase which can dephosphorylate critical regulator molecules (proteins?) involved in the eventual activation of ABA-responsive low-temperature genes through interactions with ABREs (abscisic acid responsive elements) in their upstream promoter sequences. Low-temperature exposure can also lead to elevated cellular calcium levels, independently of ABA, and although the mechanisms involved are not understood, a temperature-sensitive 'sensor' may exist as a plasma membrane component*

of their products. Some have been characterized and include antifreeze proteins and osmotin, which are also induced under conditions of drought. A recent discovery is an uncoupling protein from potatoes with properties akin to those in mammalian cells. Uncoupling proteins dissipate the proton gradient across inner mitochondrial membranes to produce heat rather than ATP, and in plants they may be involved in thermoregulatory mechanisms. Other products of LTR genes are, however, more difficult to explain in terms of acclimation mechanisms and include alcohol dehydrogenase, as well as proteins that normally appear in late plant embryogenesis, RNA binding proteins and elongation factor 1a (see Table 9.3). A coherent explanation of the biochemical basis of low-temperature acclimation in plants remains a future research challenge.

9.3 Water Stress

Environmental conditions of high salinity or drought pose considerable threats to a variety of organisms. Increases in environmental osmolality threaten cells with dehydration,

TABLE 9.3 *Some products of plant LTR genes*

Product	Other inducers
Antifreeze proteins	ABA, drought
Alcohol dehydrogenase	Drought, hypoxia
Osmotin	ABA, drought
Late embryogenesis proteins	ABA, drought
Protein synthesis elongation factor-la	
Lipid transfer protein	ABA, drought
RNA-binding protein	ABA
Uncoupling protein	
Fatty acid desaturase (cyanobacteria)	

and the productivity and quality of many commercially grown agronomic and horticultural crops are often adversely affected by high soil salinity or drought. Salt accumulation in soils caused by irrigation may ultimately inhibit the sustainability of agriculture. Both plants and bacteria, however, can respond to water stress by varying their cytoplasmic composition, through the synthesis and accumulation of compatible solutes.

9.3.1 BACTERIA

In order to reduce solute potential, bacteria can exclude salts and accumulate organic uncharged 'compatible' solutes through the activation of allosterically modulated transport proteins. These solutes are referred to as compatible because they can accumulate to high cellular concentration but do not interfere with cellular metabolism. Commonly occurring compatible compounds, many of which are also used by plants, algae and animal cells (Figure 9.6), include non-reducing sugars (e.g. trehalose), polyols (e.g. glycerol, arabitol, pinitol), amino acids (e.g. glutamate, proline) and amino acid derivatives (e.g. glycine betaine).

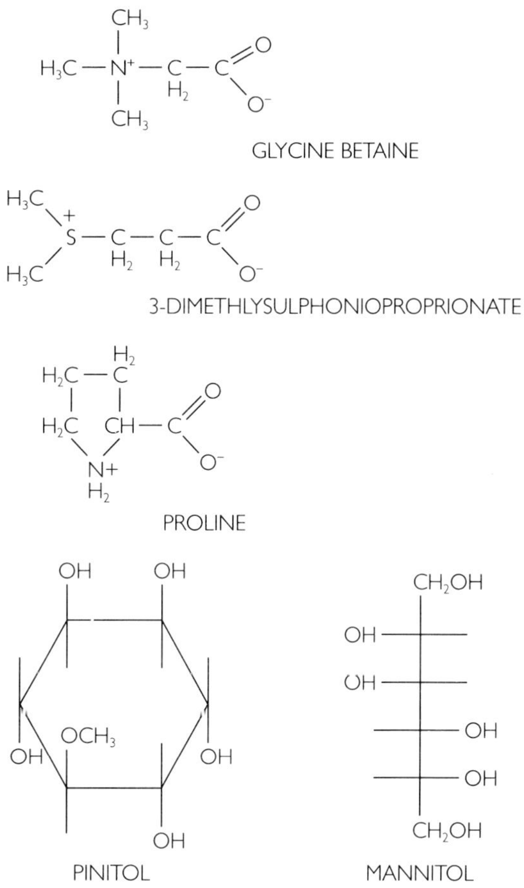

FIGURE 9.6 *Some compatible solutes that function as osmoprotectants in plants and bacteria*

This osmoadaptive capacity is based on the use of redundant pathways for the synthesis and accumulation of these cytoplasmic solutes. In response to osmotic down-shifts, rapidly activated solute efflux mechanisms (e.g. for proline, betaine and potassium) are believed to mediate rapid reductions in cytoplasmic osmolality.

Osmosensing mechanisms are required not only to detect the effects of osmotic shifts that might affect critical water-activity dependent cellular characteristics, but also to trigger gene responses. These mechanisms could include the detection of

- the hydrostatic pressure of the cytoplasm by a baroreceptor
- the pressure differential across the cytoplasmic membrane and the cell wall by a turgor pressure sensor
- the tangential strain within the cytoplasmic membrane or cell wall by stretch receptors
- cytoplasmic and periplasmic or extracellular water activity by chemoreceptors.

In *E. coli,* a two-component system appears to be involved in sensing osmotic change and activating osmotic responses. *EnvZ* is a two-component histidine kinase which functions as 'osmosensor' or 'sensory kinase'. It is activated by autophosphorylation at a histidine residue under hyperosmotic conditions and then phosphorylates an aspartate residue of the OmpR protein, which functions as a response regulator. Phosphorylated OmpR functions as a transcription factor to up-regulate the *OmpC* gene and down-regulate the *OmpF* gene. Both these genes encode proteins of the bacterial outer membrane which in combination regulate turgor pressure. Another operon, induced by increased osmotic pressure, is *proU* which encodes a transport system for

the osmoprotectant glycine betaine. Changes in DNA supercoiling mediated by an array of DNA-binding proteins and changes in cytoplasmic potassium concentrations have been invoked in the up-regulation of *proU* expression, but no unambiguous osmotic signal has yet been identified.

9.3.2 PLANTS

Like bacteria, plants can synthesize and accumulate compatible solutes in response to salt and drought stresses (Figure 9.6). These include sugar alcohols (e.g. mannitol), quaternary ammonium compounds (e.g. glycine betaine), proline and tertiary sulphonic compounds (e.g. 3-dimethylsulphonioproprionate).

Glycine betaine accumulates specifically in the cytoplasmic compartments of plant cells and its enzymic synthesis involves the two-step oxidation of choline. The first step involves the chloroplast enzyme *choline monooxygenase* which requires magnesium, molecular oxygen and reduced ferrodoxin. The second step involves *betaine aldehyde dehydrogenase*, also located in chloroplasts, which has a requirement for NAD. The activity of both enzymes is greatly increased (three- to seven-fold) in plants exposed to high NaCl concentrations (e.g. 200–400 mM). In the case of choline monooxgenase the increased activity appears to be controlled at the level of gene transcription.

In vascular plants, *mannitol* is synthesized mainly in mature leaves from mannose-6-phosphate through the action of an *NADPH-dependent mannose-6-phosphate reductase* that catalyzes the conversion of mannose-6-phosphate to mannose-1-phosphate, followed by dephosphorylation by a phosphatase. The accumulation of mannitol in salt-stressed plants paradoxically appears to involve the down-regulation of its metabolic use. Mannitol is directly oxidized to mannose by an *NAD-dependent mannitol dehydrogenase*. Mannose can then enter the tricarboxylic acid cycle for use as a carbon and energy source.

In the synthesis of the osmoprotectant *3-dimethylsulphonioproprionate* (DMSP), a cytosolic methyltransferase first converts methionine to S-methylmethionine (SMM). This is then imported into chloroplasts, where deamination and decarboxylation reactions yield DMSP. Although exposure of plants to conditions of high salinity causes accumulation of both DMSP and its precursor SMM, little is known of the regulation of the enzymes involved in their synthesis.

9.3.3 WATER-RESPONSIVE GENES IN PLANTS

Recent reports suggest that a two-component histidine kinase, analagous to the proposed bacterial osmosensor, may exist in plants. Moreover, a variety of genes have been reported to respond to water stress in various species. These can be classified into two groups.

The first group encodes proteins that probably function collectively in water stress tolerance, and includes:

■ water channel proteins involved in the movement of water through membranes, enzymes required for the biosynthesis of various osmoprotectants (e.g. glycine betaine, DMSP, proline)

■ proteins that may protect macromolecules and membranes (e.g. dehydrins, late embryonic proteins, antifreeze proteins, osmotin, protein chaperones, RNA-binding proteins)

■ proteases for protein turnover (e.g. thiol proteases, Clp protease, ubiquitin)

■ detoxification enzymes (e.g. glutathione S-transferase, catalase, superoxide dismutase, ascorbate peroxidase).

The second group contains genes that encode protein factors involved in the further amplification of stress signals and the adaptation of plant cells to water-stress conditions. These include protein kinases, transcription factors and phospholipase C.

Many water-stress inducible genes also respond to low-temperature stress, as well as to the plant growth regulator *abscisic acid (ABA)*. Analysis of mutants has shown that both ABA-dependent and ABA-independent pathways are involved in the water-stress response (Figure 9.7). Crucially, the levels of endogenous ABA increase significantly in many plants under drought and high salinity conditions. Examination of the expression of genes related to the ABA-dependent pathways has revealed that several of these genes require prior protein synthesis for their induction by ABA. Thus at least two independent pathways exist between the production of endogenous ABA and gene expression during water-stress responses, one requiring protein biosynthesis and the other not. In the first of these pathways, ABA initially stimulates the expression of genes for transcription factors analogous to the Myc and Myb factors of mammalian cells. These factors then activate specific target genes which do not contain *abscisic acid responsive elements (ABREs)*, in their upstream promoters. In contrast, in the pathway that does not require prior protein synthesis, the responsive genes do have upstream ABREs which contain a core *G-box sequence, CACGTG*. This serves as a binding site for G-box binding proteins, which contain a basic region adjacent to a leucine-zipper motif, and which contribute to the transcriptional activation of responsive genes in the second pathway.

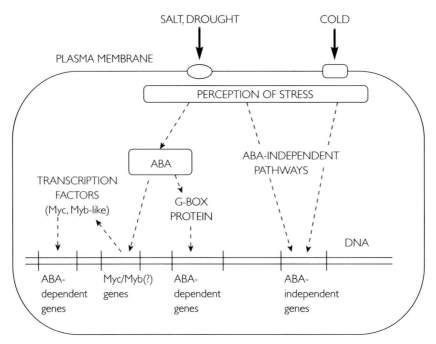

FIGURE 9.7 *A schematic representation of various signal transduction pathways between perception of water stress in plants and gene expression for stress tolerance*

9.4 *Toxic Metal Stress*

Although a number of metals are nutritionally important, many can also be cytotoxic. Some, like zinc, act as essential cofactors for a wide variety of metalloproteins and enzymes, and are not normally toxic except at extremely high concentrations. Others, like copper, are essential as enzyme cofactors but are also extremely potent cellular toxins. In contrast, metals such as cadmium are highly toxic and without nutritional value.

In biological systems, nutritionally important metal ions normally function to activate biomolecules through coordination. For example, this may be to create a structural scaffold for the establishment of proper protein conformation. Coordination of metal ions also generates more reactive electrophiles or nucleophiles fulfilling catalytic functions of enzymes. A problem is that many metal ions resemble one another, for example in terms of ionic radii and stereochemical demands, due to similarities in electronic configurations. This can cause severe problems for certain biological systems. Metal ions of high affinity such as Cd^{2+} and Pb^{2+} can compete with, or displace, weakly bound essential metal ions such as Ca^{2+} and Mg^{2+} and so inhibit the activity of certain vital metalloenzymes. Pb^{2+} at low concentrations can substitute for Ca^{2+} in intracellular second messenger signal transduction mechanisms, with deleterious effects. The functioning of certain DNA-binding proteins that regulate gene transcription is dependent on the presence of bound Zn^{2+}, and if this is replaced by Cu^{2+}, subtle transcriptional control mechanisms in animal cells can be severely disrupted. Metal ions such as Fe^{2+} and Cu^{2+} can also be hazardous for cells as they can bring about the formation of hydroxyl radicals through Fenton-type reactions (see Chapter 4). These free radicals are highly damaging towards cell components such as DNA, lipids and proteins. Thus, even metal ions that have essential cellular roles can be toxic at abnormally elevated concentrations. In microorganisms, plants and animals there are nevertheless systems that sense toxic levels of metal ions and activate the transcription of genes that encode proteins that have a protective role.

9.4.1 BACTERIAL METAL RESISTANCE

When it comes to coping with metals in the environment, bacteria are in the front line. In fact, the ease with which they tolerate high concentrations of potentially toxic metal ions is remarkable, and interest in the mechanisms involved have been heightened by increasing environmental pollution. It is now clear that plasmids within bacteria can carry resistance not only to antibiotics but also to compounds of a variety of toxic elements including antimony, arsenic, boron, cadmium, chromium, copper, mercury (and organic mercurials), lead, nickel, silver, tellurium and zinc. The plasmid loci encoding resistances to mercury, cadmium, copper and arsenic have been particularly well studied.

Mercury

Study of resistance to mercury (Hg^{2+}) demonstrated that Hg^{2+} is converted to the less toxic metallic mercury (Hg^{0}) by resistant bacterial cells. The process is inducible by exposure of cells to sub-lethal concentrations of mercury compounds and, in Gram-negative bacteria, involves the inducible plasmid-borne mercury resistance (*mer*) operon of around 3700 base pairs. This contains genes that encode a highly hydrophobic transmembrane protein (*merT*), a periplasmic Hg-binding protein (*merP*), a mercuric ion reductase (*merA*) which serves as a detoxification enzyme, and an organomercurial lyase (*merB*) which cleaves the C–Hg bond in compounds such as methyl- and phenyl-Hg. Because the reductase is intracellular, Hg^{2+} has first to be taken into cells and the specific uptake system comprises the transmembrane protein and the periplasmic Hg-binding protein. The mercuric ion reductase is a flavoprotein which requires NADPH as cofactor. The rate of synthesis of these proteins is controlled by at least one regulatory protein, the product of the *merR* gene (Figure 9.8). This is a DNA-binding protein that can act on the operator/promoter region and has regulatory functions activated in the presence of Hg^{2+} or phenyl-mercury.

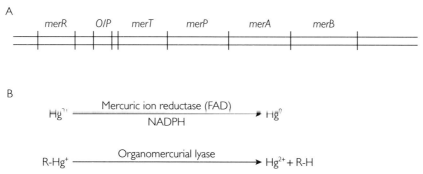

FIGURE 9.8 *(A) A schematic representation of the plasmid-borne* mer *operon containing genes involved in resistance to mercury and organomercurials in Gram-negative bacteria (*merR*, regulatory protein; O/P, operator/promoter region;* merT, *transmembrane protein,* merP, *mercury-binding protein;* merA, *mercuric ion reductase;* merB, *organomercurial lyase); (B) the biochemical basis of mercury resistance*

Arsenic

Although arsenate can be taken up by bacterial cells through the phosphate uptake pathway, in *S. aureus* and *E. coli* resistance is achieved through an ATP-dependent efflux system. This is encoded in a plasmid-borne operon, *arsABC*, regulated by a linked *arsR* gene. The *arsA* gene specifies an ATP-binding protein with ATPase activity which interacts with an inner membrane protein coded by the *arsB* gene to form the arsenate efflux pump. To pump out arsenate, the *arsC* gene product is required to be added to the complex.

Copper

Copper resistance, which involves sequestration, is mainly encoded by four linked plasmid genes termed *pcoARBC*. The *pcoR* gene specifies a DNA-binding protein involved in the induction of copper resistance. PcoC is a cytoplasmic copper transport or storage protein, while PcoA and PcoB are necessary for chemical modification of copper and its export, although they also require the participation of two chromosomal genes *cutC* and *cutD* for these processes.

Cadmium

Various mechanisms of resistance to cadmium have been identified. Two of these require the genes *cadA* and *cadB* (which also encode resistance to Zn^{2+}). *CadA* specifies an energy-dependent efflux system with the properties of a cation translocating ATPase. *CadB* is an inducible cadmium-binding protein.

9.4.2 METAL-REGULATED TRANSCRIPTION IN EUKARYOTES

Metallothioneins

In higher eukaryotes the metallothionein genes are probably the best understood metal-regulated transcription units. *Metallothioneins* are low molecular weight cysteine-rich metal-binding proteins found in a wide range of eukaryotes and are believed to play several roles in metal homeostasis.

In mammals, metallothioneins (Class I) comprise approximately 61 amino acids. While they have no aromatic amino acids, 20 cysteine residues are located in invariant positions in Cys–Xaa–Cys motifs (where Xaa is an amino acid other than Cys). These can coordinate seven divalent metal ions (or 12 monovalent ions) in two distinct metal clusters. Class II metallothioneins are found in certain yeasts, nematodes and cyanobacteria. These have a Cys–Cys and a Cys–Xaa–Xaa–Cys within the N-terminal domain. In plants the only protein that can be classed as a metallothionein is the E_C protein of wheat. This protein can associate with Zn^{2+} in a stoichiometric ratio of 5:1 and although it has Cys–Xaa–Cys motifs, its designation as a Class II metallothionein is due to additional properties. Plants and certain fungi also contain poly(γ-glutamyl-cysteinyl)glycine which can bind Cd^{2+}. These polypeptides and other γ-glutamyl isopeptides in which glycine is either absent or substituted with β-alanine are sometimes referred to as Class III metallothioneins.

Metallothionein gene transcription

In higher eukaryotes, the metallothionein (MT) genes are perhaps the best understood

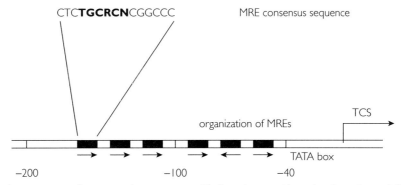

CTCTGCRCNCGGCCC   MRE consensus sequence

organization of MREs

TCS

TATA box

−200   −100   −40

FIGURE 9.9 *The consensus metal-responsive element (MRE) of higher eukaryotes (the nucleotide residues in bold type are those that are critical for function) and the organization of MREs, within the mouse metallothionein-I (mMT-I) gene promoter*

examples of metal-regulated transcription units. The role for metallothioneins in protecting organisms from the toxic effects of metals correlates with the ability of a number of metals such as zinc, copper, cadmium and lead to activate promptly MT gene transcription. In the upstream promoter region of the intensively studied mouse metallothionein-I gene (*mMT-I*), multiple sequence elements are present that contribute to the overall extent of the metal induced transcription. These *metal-responsive elements (MREs)* are present in multiple copies in all MT promoters thus far examined. They comprise a series of 13–15 imperfect repeat sequences found in both orientations (Figure 9.9).

For the activation of *MT* genes, the residues in the core sequence of an MRE are more critical than the GC-rich flanking sequences (Figure 9.9), and MRE sequences are conserved in humans, mice, sea urchins, fish, flies, and other higher eukaryotes. While a single MRE is capable of responding to a range of metals such as cadmium, zinc and copper, it is believed that one or more metal-responsive transcription factors (MRTFs) are necessary for interaction with MREs. Cadmium-dependent protein–DNA interactions have been detected in all the MREs present in the rat *MT-I* gene

promoter (in general cadmium appears to be the most potent metal for *MT* gene induction in higher eukaryotes). It is possible that multiple MRE binding transcription factors exist with metal-dependent activities, but each with distinct metal specificities. However, the possibility of a single MRTF within eukaryotic organisms that might sense and respond to a variety of different metals may be more likely, in light of recent experimental evidence. In addition, there is evidence for the existence of an inhibitory protein associated with such an MRTF, which is released in the presence of metals. Despite this, there has been limited progress towards understanding the regulation of *MT* genes in higher eukaryotes, aspects such as the transport of metal into cells and their nuclei and the overall homeostasis of metals.

Information on metal-regulated gene transcription in yeasts is also available. Copper detoxification in *Saccharomyces cerevisiae* is carried out by metallothioneins encoded by the *CUP 1* gene, which shares some limited homology with mammalian *MT* genes. Gene deletion studies have shown that the *CUP 1* gene product, although not essential for normal growth, is critical when yeast cells are exposed to high concentrations of copper.

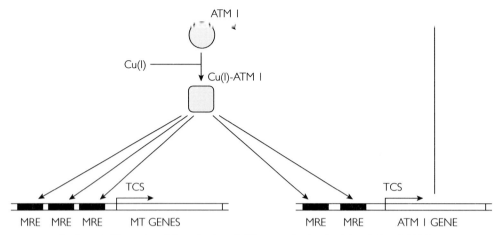

FIGURE 9.10 *A schematic illustration of the activation of MT gene activation involving Cu⁺ and ATM 1*

Unlike mammalian *MT* genes, *CUP 1* is only induced by Cu^+ and Ag^+. Induction by copper is achieved through one or more positive-acting copper-responsive regulatory factors. *CUP 1* gene transcription is also subject to negative autoregulation, as basal levels of the metallothionein *CUP 1* gene product can sequester copper from the positive-acting regulatory factors. One of these regulatory factors is encoded in the *ACE 1* gene and is itself a cysteine-rich Cu(I)-binding protein. It appears to function both as a copper sensor and as a sequence-specific DNA-binding transcription factor. The sensing capability is achieved by Cu(I) coordination through cysteinyl thiolates interdigitated within the Cu^+-activated DNA-binding domain. The coordination of Cu^+ by ACE 1 is transduced into a biological signal by altering the ACE 1 DNA binding domain conformation which triggers sequence-specific binding of monomeric ACE 1 to four MREs in the *CUP 1* gene promoter. In comparative terms, Ag^+ is a less potent inducer of the *CUP 1* gene. This may be due to its larger ionic radius compared with Cu^+. It may be that an ACE 1 complexed with Ag^+, rather than Cu^{2+}, has a lesser affinity for MRE sequences.

In contrast, the opportunistic pathogenic yeast *Candida glabrata* has a large family of *MT* genes comprising a single *MT-I* gene, a tandemly amplified *MT-IIa* gene, an unlinked *MT-IIb* gene plus additional genes encoding MT isoforms. While these genes can be shown to function in copper detoxification, their transcription again is only up-regulated by Cu^+ or Ag^+. As was the case for *S. cerevisiae*, a Cu-sensing transcription-activating molecule can be detected within *C. glabrata*. The synthesis of this protein, AMT 1, is autoregulated in response to copper and it exhibits a considerable degree of sequence homology with ACE 1 described above. Besides activating its own *ATM 1* gene, it also activates the whole family of *C. glabrata MT* genes. Like ACE 1, ATM 1 binds Cu^+ through its cysteinyl thiolates which alters its DNA-binding domain conformation and triggers the sequence-specific interactions of ATM 1 with the MREs of the *MT* and *ATM 1* genes (Figure 9.10). The metallothioneins produced as a result of the activation of these genes then chelate free intracellular copper, thus preventing cytotoxicity.

In plants, information describing the expression of metallothionein-like genes is very limited. Plants certainly contain diverse met-allothioneins with the potential to perform discrete functions in the metabolism of various metal ions. For example, the E_C protein from wheat appears to be involved in the endogenous regulation of Zn^{2+} metabolism during normal plant embryo development, while type I metallothioneins are likely to be involved in the sequestration of copper in roots. There, they are expressed at high levels, even in nutrient conditions devoid of supra-optimal levels of trace-metal ions. Cadmium detoxification, in contrast, seems to involve non-gene-coded poly(γ-glutamyl-cysteinyl)glycine-type polypeptides, the so-called type III metallothioneins.

9.5 *Oxidative Stress*

Aerobic metabolism confers great energy benefits to the organisms in which it evolved. A constant hazard, however, is the potential for oxidative damage arising from by-products of respiration and photosynthesis, such as superoxide and hydrogen peroxide, to cellular macromolecules such as DNA, protein and lipids. As outlined in Chapter 4, these particular oxygen species can give rise to extremely damaging hydroxyl free radicals in Fenton-type reactions. Bacteria, as well as surrounding mammalian tissues at sites of inflammation, can also suffer excessive exposure to superoxide and hydrogen peroxide released from stimulated macrophages and neutrophils. Oxidants are also generated in mammalian tissues exposed to environmental agents such as asbestos fibres or to the components of cigarette smoke (see Chapter 2). Moreover, redox-active chemicals, such as the herbicide paraquat, can generate excessive levels of superoxide within plant and animal cells, as well as bacteria, as a result of intracellular oxidation–reduction reactions. In general, oxygen itself, at concentrations greater than that of normal air (i.e. 21%), has notable toxic effects on animals. Damage to mitochondria of heart and kidneys is observed, as well as injury to lung tissue and the impairment of erythroid development in bone marrow. Increased chromosomal aberrations are also evident, and cell proliferation is generally inhibited. These toxic effects of oxygen appear to stem from the resulting formation within cells of abnormal levels of reactive oxygen species such as superoxide, hydrogen peroxide and hydroxyl radicals, which, in turn, can cause severe oxidative damage to vital cellular components.

Organisms can combat the threat of such oxidative stress with various antioxidants and antioxidant defence enzymes. For example, superoxide dismutase, catalase and glutathione are almost universally present in aerobic organisms and limit the oxidative decay of cellular structures and DNA. The expression of these activities can be modulated in response to various environmental cues and the coordinated expression of multiple defence functions comprises the *oxidative stress response* in many organisms.

9.5.1 THE ADAPTIVE RESPONSE TO OXIDATIVE STRESS IN BACTERIA

The *oxyR* regulon

Prior exposure of *E. coli* to low (i.e. micromolar) concentrations of hydrogen peroxide confers tolerance to subsequent exposure to higher (millimolar) concentrations that would be lethal under normal circumstances. The acquisition of this tolerance is accompanied by the elevated expression of some 30 proteins, of which four have identified antioxidant defence capabilities. In *S. typhimurium* and

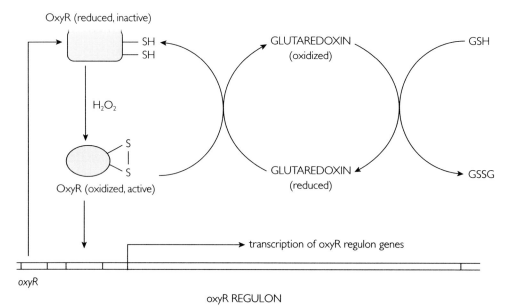

FIGURE 9.11 *The role of oxidized and reduced OxyR protein in the activation of the oxyR regulon by hydrogen peroxide (GSH, reduced glutathione; GSSG, oxidized glutathione (glutathione disulphide))*

E. coli one regulon, *oxyR*, is responsible for regulating the expression of nine of the inducible proteins including *catalase, superoxide dismutase, alkyl hydroperoxide reductase* and *glutathione reductase*, which are increased two- to 40-fold. The *oxyR* gene encodes a 34.4 kDa protein, OxyR, which functions as a positive regulator. During conditions of oxidative stress, a direct oxidation of the OxyR protein causes a conformational change whereby OxyR transduces an oxidative stress signal to the transcriptional apparatus and thereby activates the expression of at least nine other proteins. At a molecular level the activation of OxyR is a reversible process involving the formation of an internal disulphide between cysteine residues at positions 199 and 208. Within the cell OxyR is placed between two opposing pathways (Figure 9.11). In one, it is activated by hydrogen peroxide-mediated oxidation to give the disulphide, and in the other it can be inactivated by reduction through the thiol donors glutaredoxin and GSH. These pathways permit a dynamic response in which OxyR is inactivated rapidly after the withdrawal of the oxidative stress (Figure 9.11).

The *soxRS* regulon

Pretreatment of *E. coli* with superoxide-generating agents (e.g. the redox active herbicide, paraquat) increases the tolerance of the bacteria to paraquat toxicity. This resistance is accompanied by the increased expression of around 40 proteins including *superoxide dismutase, glucose-6-phosphate dehydrogenase* and *endonuclease IV*, as well as heat-shock proteins such as *GroEL, GroES* and *DnaK*.

Of this group of 40 inducible proteins, 12–14 are encoded by genes of the *soxRS* regulon which can be activated by superoxide-generating agents, as well as nitric oxide. The regulon contains two regulatory genes, *soxS* and *soxR*. The *soxR* gene controls the expression of the *soxS* gene which in turn yields the SoxS protein (15 kDa), which subsequently acts to activate the other genes of the regulon

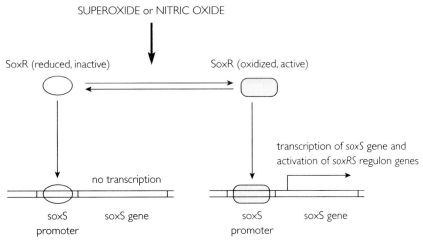

FIGURE 9.12 *The role of oxidized and reduced SoxR protein in the regulation of soxS gene transcription and the activation of soxRS regulon genes*

(Figure 9.12). The SoxR protein (17 kDa), which acts as a homodimer, contains non-haem iron that functions to alter the structure of the complex formed between the SoxR dimer and the upstream promoter sequence of the *soxS* gene so as to stimulate transcriptional initiation by σ^{70}-containing RNA polymerases (Figure 9.12). The iron in SoxR is organized as pair of [2Fe–2S] clusters, one in each subunit of the homodimer. These FeS centres are redox active, and present evidence indicates that the transcriptional activity of SoxR is regulated by oxidation and reduction. As indicated in Figure 9.12, inactive SoxR binds to the *soxS* gene promoter but does not stimulate transcription. Superoxide, or nitric oxide, converts SoxR to the oxidized active form that stimulates transcription of the *soxS* gene.

9.5.2 EUKARYOTIC RESPONSES TO OXIDATIVE STRESS

An emerging feature of the regulation of oxidative responses in yeasts and higher eukaryotes is that these are much more complex than those observed in prokaryotes. No homologues of the SoxR, SoxS or OxyR proteins have yet been detected in eukaryotes. Moreover, eukaryotes manifest considerable overlap between oxidative stress responses and other stress responses, such as resistance to metals and heat shock. In view of the potential for metal ions, such as Fe^{2+} and Cu^{2+}, to generate reactive free radicals through Fenton-type reactions (see Chapter 4), it may make biological sense to have coordinated responses to metals and oxidant stress.

In the yeast, *Saccharomyces cerevisiae*, the *yap1* gene plays a role in the adaptive response to hydrogen peroxide. At the molecular level its product, *Yap1 protein*, appears to regulate the expression of several enzymes with anti-oxidant activity such as *superoxide dismutase, glucose-6-phosphate dehydrogenase, thioredoxin* and *glutathione reductase*. In support of this, putative Yap1-binding sites are found in the promoter regions of genes encoding these particular proteins. Another protein, *Yap2*, is also involved in the adaptive response to hydrogen peroxide although no target genes have been identified. *Yap2* is also involved in conferring resistance on yeasts to the toxic effects of cadmium.

In mammalian cells there is also evidence of adaptive responses to hydrogen peroxide. In addition to the heat-shock genes, genes such as *c-jun* and *c-fos* as well as the genes encoding haem oxygenase and a protein tyrosine kinase are induced by oxidants. Surprisingly, evidence for increased transcription of antioxidant defence enzyme genes, such as catalase, superoxide dismutase and glutathione peroxidase, under these conditions is inconsistent. While it has been pointed out in section 9.1 that hydrogen peroxide is an effective inducer of heat-shock protein synthesis in mammalian cells, due to its ability to increase the intracellular levels of non-native proteins, probably by oxidation, less is known about the regulation of the other inducible genes mentioned. However, two classes of mammalian transcription factor, *NF-κB* and *AP-1*, may play central roles.

NF-κB is a heterodimeric DNA-binding protein made up of 50 kDa and 65 kDa subunits important to a number of immune and inflammatory responses. Normally NF-κB forms a complex with an inhibitor subunit IκB located in the cytoplasm. Upon activation, the inhibitor is released from NF-κB, and the transcription factor is translocated to the nucleus, where it induces the transcription of target genes. Among other things, both the activation and DNA-binding of NF-κB are strongly induced by hydrogen peroxide. The subsequent observation that activation is inhibited by antioxidants has led to the suggestion that reactive oxygen species are a common signal for the activation of NF-κB. AP-1 is another transcription factor whose ability to bind to the promoter sequences of a number of target genes is dependent on its redox state. AP-1 is a dimer of the protein products of the *c-fos* and *c-jun* genes, and oxidation of conserved cysteine residues in the DNA-binding domains leads to reduced binding ability.

In the case of plants it appears that both exogenous and endogenous levels of hydrogen peroxide can affect catalase activity. Indeed, hydrogen peroxide may play a significant signalling role in influencing the tissue-specific and temporal expression of different catalase genes in maize. Unlike animals, plants contain multiple enzymatic forms of catalase, and in addition to scavenging hydrogen peroxide, this array of catalases may have an important role in normal plant developmental programmes as well as in responses to environmental insults. However, at present there is little detailed knowledge of any transcriptional regulation mechanisms involved. There is also evidence that indicates a role for superoxide dismutases in protecting plants from the damage associated with oxidative stress. Increased expression of superoxide dismutase genes has been observed in plants following exposure to superoxide-generating agents such as paraquat. However, as was the case for the plant catalase genes, the transcriptional regulation mechanisms remain obscure. None the less, levels of superoxide dismutase activity have been correlated with mechanisms of herbicide tolerance, and overexpression of superoxide dismutase genes in transgenic plants increases their resistance to paraquat. As was the case for hydrogen peroxide, there is also the possibility that, in plants, superoxide may play a role in signal transduction and thus its modulation by superoxide dismutases may be an important additional role for these antioxidant defence enzymes. Other plant enzymes with roles in protection from oxidative stress are ascorbate peroxidases and glutathione reductases, but at present there are conflicting data with regard to any regulation of their genes in the face of environmentally imposed oxidative stress. However, the plant growth regulator, ethylene, which is known to be liberated from plants exposed to various environmental pollutants, can elevate levels of ascorbate peroxidase enzyme activity which correlate with protection of plants against hydrogen peroxide, paraquat and ozone.

9.6 *Hypoxic and Anoxic Stress*

Perhaps the ultimate extreme in environmental tolerance for the majority of terrestrial species is oxygen deprivation. In humans and other mammals this can be a hazard for vital tissues that suffer either a partial (hypoxic) or total (anoxic) deficiency in situations where blood supplies are impaired. For higher plants, flooded or waterlogged habitats that carry the risk of being anoxic are potentially detrimental. Deprivation of oxygen in mammalian and plant cells rapidly reduces the supply of ATP, and while a period of anoxia can be survived, it can cause tissue death if prolonged.

9.6.1 PLANTS

Most plant species that are able to grow in flooded or waterlogged soils are able to surmount the hazards of anoxia by virtue of highly developed aeration systems. However, there are examples of plant tissues such as germinating seeds, root tips, rhizomes and stolons, the cambial zone in tree trunks and developing tree buds, where the demand for oxygen is often greater than supply and the tissues suffer either a hypoxic or an anoxic stress.

However, an additional type of stress comes when, after a period of oxygen deprivation, supplies of oxygen from the air once more become available. At this time peroxidative damage to tissue components is a severe risk from reactive by-products of oxygen metabolism such as superoxide and hydrogen peroxide. Unfortunately, the ability to combat oxidative damage is frequently at a minimum after a period of oxygen deprivation. This second type of stress is often referred to as *post-anoxic injury.* In many respects it is similar to the injury that can be suffered by human heart muscle tissue following a heart attack, in which oxygen supply is briefly arrested and then restored. The problem is referred to as post-ischaemic or reperfusion injury, and is due to peroxidative damage to muscle tissue, arising from the generation of reactive forms of oxygen when air returns to the heart.

In maize roots, although anoxia rapidly inhibits general protein synthesis, a select group of *anaerobic proteins (ANPs)* continue to be made. Most of the 20 or so ANPs thus far identified appear to be enzymes of glycolysis and fermentation. In general, oxygen deficiency strongly stimulates the transcription of genes for ANPs, although not all are translated, which may indicate that maintenance of glycolysis and fermentation pathways has a high priority under conditions of anoxic stress. In soybean roots, which are less tolerant of anoxia than maize, only four proteins are evident, one of which is alcohol dehydrogenase. It may be that alcohol dehydrogenase serves to regenerate metabolic oxidizing power in the shape of NAD and thus drive glycolysis. Alcohol dehydrogenase appears necessary to survive anaerobiosis, but levels of alcohol dehydrogenase are greatly in excess relative to the rate of fermentation in anoxic roots, and above a certain threshold, levels do not correlate with ability to survive anoxia. Other studies, however, have shown that levels of pyruvate decarboxylase more closely relate to fermentation rates and may have a central role in regulating ethanol fermentation. Several genes encoding ANPs share a common consensus sequence motif in their promoter regions, termed the *anaerobic responsive element (ARE).* These AREs may function in the coordinate induction of plant genes under anoxia, and

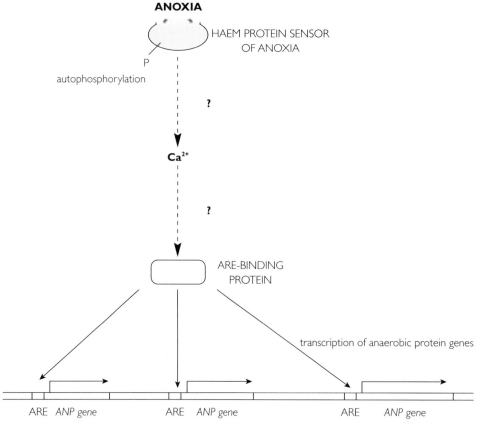

ANOXIA

HAEM PROTEIN SENSOR
OF ANOXIA

P
autophosphorylation

?

Ca²⁺

?

ARE-BINDING
PROTEIN

transcription of anaerobic protein genes

ARE ANP gene   ARE ANP gene   ARE   ANP gene

FIGURE 9.13 *A schematic diagram to illustrate possible signal pathways whereby anaerobic protein (ANP) genes might be activated during anoxia (ARE, anaerobic responsive element)*

nuclear proteins have been identified which can bind to AREs.

An important question still to be addressed is how the signal transduction systems responsible for ANP induction perceive oxygen deficiency. Work with bacterial systems has identified sensor response regulatory systems. In these cases the sensor appears to be an oxygen-binding haem protein which responds by autophosphorylation as the initial step in the signal tranduction process. It may be that haem also contributes to the sensing of oxygen concentrations in plants, and there are indications that transient changes in the cytosolic concentration of Ca²⁺ may also be part of the signal transduction pathways (Figure 9.13).

As already mentioned, another hazard for plants is post-anoxic injury. This can occur when water tables drop and plants are once again exposed to air. During the period of recovery from anoxia it has been observed that there are increases in the tissue levels of *superoxide dismutase* in certain plants. Thus a possible protective role for superoxide dismutases in the post-anoxic recovery processes is suggested.

9.6.2 MAMMALIAN TISSUES

In cells of mammalian tissues, a notable feature of hypoxia is the induction of the *p53* gene. Although the mechanism of induction is not known, it has been suggested that mammalian cells may detect low oxygen levels through a haem-containing sensor protein which leads to the activation of a *hypoxia-inducible factor 1α (HIF-1α)* that stabilizes p53 and permits the stimulation of the transcription of several genes associated with hypoxia, sometimes referred to as the *oxygen-regulated protein genes*. In Chinese hamster cells the oxygen-regulated proteins include two proteins which are also induced by glucose deprivation (*glucose-regulated proteins*, GPR 94 *and* GPR 78). Hypoxia also induces genes associated with drug detoxification, including those for *NADPH: quinone oxidoreductase* and *γ-glutamyl-cysteinyl-synthetase*, as well as two growth arrest and DNA damage inducible (GADD) genes (*gadd45* and *gadd153*) together with *haem oxygenase-1 (HO-1)*. HO-1 serves an important physiological function in regulating, by catabolism, the intracellular concentration of haem in mammalian tissues. HO-1 is also notable for its induction by a wide spectrum of physical and chemical agents which include its substrate, haem, as well as heavy metals, heat and ultraviolet A radiation (see below). It can also be induced by oxidants, and its elevated expression in mammalian ischaemic tissues after reperfusion may well be a consequence of induction by reactive oxygen species.

9.7 *Sunlight and Ultraviolet Radiation Stress*

9.7.1 PHOTOOXIDATIVE STRESS IN PLANTS

The leaves of plants normally intercept light and, in photosynthesis, transform it into energy (ATP) and reductant (NADPH), which are used primarily for the reduction of carbon dioxide. However, if exposed to levels of light in excess of those required for generation of reductants, the antennae of the photosynthetic apparatus can transfer this excess excitation energy to ground state oxygen with the resultant generation of *singlet oxygen* (see Chapter 4). In addition, electron transport systems driven by light can divert electrons from $NADP^+$ to O_2, thus generating superoxide radicals. Such production of reactive oxygen species in response to excessive light is referred to as *photooxidative stress*, and can lead to oxidative damage of cellular components such as the photosynthetic apparatus itself (*photoinhibition*). However, as pointed out in Chapter 5, plants contain a wide array of antioxidant defence systems. Although these are constitutively present in leaves, the overall protective capacity is not constant but responds to environmental factors such as light intensity. In general, plants grown at high light intensities have higher levels of antioxidant protection including *ascorbate, ascorbate peroxidase, dehydroascorbate reductase* and *glutathione reductase*. However, there is no evidence that these sytems can adjust instantaneously to rapid changes in light intensity. Other environmental conditions that limit the use of reductants produced in photosynthesis are low and high temperatures, which thus increase photosynthetic stress.

9.7.2 ULTRAVIOLET RADIATION STRESS IN PLANTS

While solar radiation of the visible spectrum necessary for photosynthesis readily penetrates the atmosphere to sea level, so also do the longer wavelengths of ultraviolet (UV) radiation. These are potentially damaging to plants and their DNA, and such damage is likely to become more ecologically significant as a consequence of reductions to the stratospheric ozone layer. However, there is a wide diversity in plant resistance to UV, and although visible symptoms of UV injury in the field are rare, experiments have been carried out under conditions intended to mimic levels of UVB that might be encoutered following a stratospheric ozone depletion. Plants respond to such conditions with the induction of enzymes associated with the synthesis of UVB protective *flavonoids* in the epidermal layers, with the resultant bronzing or reddening to leaves. The transcriptional activity of the gene for *chalcone synthase*, the first enzyme in the flavonoid biosynthetic pathway, is greatly stimulated by UV light, with calcium and calmodulin as likely components of the intracellular signal transduction system involved. Another response is the enhanced repair of UV-induced DNA pyrimidine dimers in a direct photoreactivation process involving *photolyase enzymes* (see section 6.1). UVB also affects plant membranes, giving rise to increased cell permeability and products of lipid peroxidation suggesting enhanced free radical activity and oxidative stress. In part this may be offset by a concomitant increase in extra antioxidants.

9.7.3 RESPONSES TO UV RADIATION IN BACTERIAL AND ANIMAL CELLS

UV radiation can cause both death and mutation of cells, as well as changes in cell cycle progression and alterations in membrane permeability. However, since UV interacts quite differently with cells depending on its wavelength, the stress responses elicited by the solar UVA (320–380 nm), the solar UVB (290–320 nm) and the non-solar UVC (<290 nm) radiations are best detailed separately as below.

Chronic exposure to both UVA and UVB components of sunlight is a central factor not only in photoageing of human skin but in the development of both non-melanoma and melanoma type skin cancers in humans. At present, however, a great deal of information is still required regarding the physiological and pathological significance of the many UV-inducible responses observed.

UVC

UVC is strongly absorbed by DNA and leads to the types of DNA damage described in section 2.3. In bacteria this can lead to the 'SOS response' (see section 6.7) which entails the induction of an 'error-prone' type of DNA repair mechanism, that leads to elevated levels of mutations. Although there is no direct equivalent of the SOS response in eukaryotic cells, radiation-induced DNA damage can 'signal' the induction of the transcription factor p53. As discussed in Chapter 8, p53 has a unique role in orchestrating the cellular responses to DNA damage. UVC radiation will also activate a number of other mammalian genes including those for collagenase, plasminogen activator, and certain cytokines, as well as the protooncogenes *c-fos* and *c-jun* which will result in the accumulation of the AP-1 transcription factor (dimer of Jun and Fos). This, together with the direct activation of another transcription factor, NF-κB, can potentially lead to the up-regulation of many genes which may play a protective role in the stress response to UVC. At present, however, precise information on the functioning of such UVC-activated genes remains elusive.

UVA

The interaction of UVA radiation with cells, such as skin fibroblasts, is distinct from that of UVC. While DNA and proteins do not absorb radiation at these longer wavelengths a variety of low molecular weight molecules within the cells participate in photochemical reactions to generate free radicals and other reactive oxygen species such as singlet oxygen (see Chapter 4) which can oxidatively damage cellular constituents within range, for example membrane components. Another consequence is the up-regulation of a different spectrum of genes from UVC radiation. These include the genes for the protein phosphatase CL100, phospholipase A2 and haem oxygenase-1, as well as collagenase.

The gene for the protein phosphatase CL100 is also activated by hydrogen peroxide, heat shock and exposure to certain growth factors, and its cellular targets are certain of the cellular MAP kinases (mitosis-activated protein kinases). For this reason, CL100 may be involved in the down-regulation of intracellular signalling pathways normally activated by mitogens and stress, and thus it is likely that CL100 plays a key role in the response of fibroblasts to this oxidizing component of solar UV radiation.

In contrast, the induction of phospholipase A2 is possibly required for the repair of UVA-induced membrane damage. UVA will also cause the release of interleukin-6, instrumental in the activation of the collagenase genes which may be involved in collagen breakdown that is an underlying cause of actinic elastosis, a notable feature of photo-ageing of the skin.

The haem oxygenase-1 gene is induced not only by UVA but by a wide range of other stressors including heavy metals, heat shock, anoxia, hydrogen peroxide, and thiol active compounds as well as by phorbol esters, cytokines, and its substrate haem. Indeed, the proximal promoter sequence of the *HO-1* gene is extremely complex and contains metallothionein-like metal responsive elements and heat-shock elements as well as sequence elements responsive to NF-κB, interleukin-6, phorbol ester and prostaglandins.

Possible roles for *haem oxygenase (HO)* in response to these various stresses have been speculated. Increased production of the bile pigments biliverdin and bilirubin as a consequence of HO activity has been proposed to have beneficial antioxidant properties in serum with respect to a whole organism subjected to oxidative stress. The activity of HO in organs involved in the breakdown of haem from disintegrating erythrocytes probably contributes to such a mechanism. It remains unclear, however, whether bilirubin would have any beneficial antioxidant function at the cellular level prior to its efflux into the blood. In contrast, HO could affect intracellular iron metabolism by releasing iron bound to haem, thus transiently releasing it to a low molecular weight chelatable or 'free' form which could act as a prooxidant catalyzing membrane peroxidation and/or oxidative modification of DNA. However, a direct relationship exists between levels of intracellular 'free' iron levels and the biosynthesis of the iron storage protein ferritin, which limits the potential for iron to catalyze oxidative reactions. A proposal is that free iron, including that released from haem through the action of HO, plays a key role in intracellular iron metabolism by regulating ferritin synthesis. Prior irradiation of human skin with UVA confers a transient resistance to oxidative membrane damage following subsequent UVA exposure. This resistance is associated with the induction of HO activity and ferritin synthesis. In summary, it has been proposed that ferritin has the crucial cytoprotective role, and that HO-1 induction is an intermediate in the process of ferritin production by catalyzing the release of regulatory iron from haem.

UVB

UVB radiation is absorbed by DNA and can cause DNA damage. It can also penetrate sufficiently into skin to damage the epidermis, and even the dermis, severely. In situations where human skin is unprotected, it is the most important type of UV radiation in terms of severe acute effects such as sunburn and chronic effects that include basal and squamous cell carcinomas. While UVB activates a similar set of genes to UVC, it also has UVA-like properties in terms of its oxidative properties.

9.8 Toxic Chemical Stress

Many synthetic and naturally occurring chemicals in the environment are stressful because they can cause severely adverse biological effects in the cells of plants and animals. However, some protection from certain of these potentially toxic chemicals is possible because of their ability to bring about the activation of genes in these cells that encode detoxification enzymes or transporter ATPases. The detoxification enzymes directly reduce the activity of many of these harmful compounds, or increase their water solubility as an initial step in their metabolism and secretion. On the other hand, the transporters neutralize the stressful effects of likely toxic chemicals by removing them from target cells. In contrast, certain chemicals in the environment can mimic, or disrupt, the biological effects of naturally occurring hormones. This different type of chemical stress can result in the inappropriate regulation of genes required for key processes in normal animal development, with potentially serious biological consequences.

9.8.1 ACTIVATION OF DETOXIFICATION ENZYME GENES

The adverse biological effects in animals brought about by chemicals such as polycyclic aromatic hydrocarbons and halogenated aromatic hydrocarbons (e.g. 2,3,7,8-tetrachloro-*p*-dioxin, TCCD) include tumour promotion, teratogeneisis, hormonal disregulation and immunosupression. It is believed that these diverse effects are mediated by the activation of the *aromatic hydrocarbon receptor protein (ARH)*, and the subsequent upset of cellular homeostasis. ARH is a ligand-dependent transcription factor which is normally expressed early in development and is critical for important developmental functions. Ligand activation permits it to dimerize with the *ARH nuclear translocator protein (ARNT)* which facilitates its migration from the cytosol to the cell nucleus. There the ARH–ARNT dimer functions as a transcription activator through binding to sequence elements in gene promoters referred to as *AhREs (aromatic hydrocarbon responsive elements)* or *XREs (xenobiotoic responsive elements)* with the consensus structure, *T-GCGTG* (Figure 9.14).

Particular genes with upstream XREs which are up-regulated by ARH–ARNT complexes include several detoxification enzymes mentioned in Chapter 5, such as the *cytochrome P450s (CYP1A1, CYP1A2 and CYP1B1)*, as well as *UDP–glucuronosyl transferase, glutathione S-transferase* and *aldehyde dehydrogenase*. Additionally, the ARH–ARNT complex will stimulate the transcription of the genes for interleukin-1β and plasminogen activator inhibitor II as well as the protooncogenes *c-jun* and *junD*. In summary, AHR plays a vital part not only in the regulation of xenobiotic metabolism but also in the maintenance of homeostatic functions.

The expression of other cytochrome P450 is stimulated by other inducers. For example,

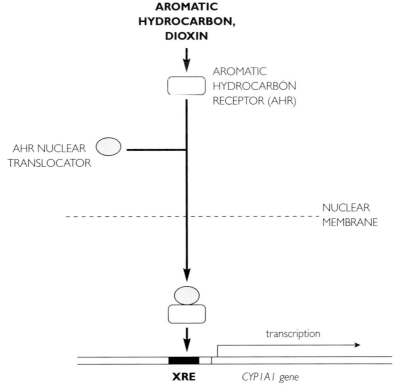

FIGURE 9.14 *A schematic diagram illustrating possible steps in the activation of the* CYP1A1 *gene by aromatic hydrocarbons or dioxins*

phenobarbitone, terpenoids and related compounds increase the transcription of genes for *CYP2B, CYP2C, CYP3* and *CYP6*, whereas glucocorticoids and peroxisome proliferators stimulate genes for *CYP3* and *CYP4*.

The genes for *glutathione S-transferases, UDP–glucuronosyl transferase* and *NAD(P)H quinone oxidoreductase (quinone reductase)* can also be induced in certain animal cells exposed to a variety of electrophilic compounds and phenolic antioxidants (e.g. tert-butylhydroquinone, 1,2-dithiole-3-thionone, isothionates, diphenols, thiocarbamates). In many cases this involves the activation of upstream promoter *electrophile-responsive elements/antioxidant-responsive elements (EpRE/AREs)* that consist of two non-overlapping consensus sequences *GTGACA(A/T)(A/T)GCNNGC* which act as binding sites for the AP-1 transcription factor

complex. Many differences, however, have been reported regarding the details of the regulatory mechanisms between different animal species and individual genes.

The possibility of induction of human *glutathione S-transferases (GSTs)* by drugs has triggered a growing interest in the possibilities of developing *chemointervention strategies* for human populations routinely exposed to carcinogens. For example, aflatoxin B_1 is a potent dietary toxin and hepatocarcinogen to which humans are commonly exposed in several areas of the world. Aflatoxin B_1 is initially metabolized in the liver by cytochrome P450 to yield aflatoxin B_1-8-9-epoxide, which preferentially attacks guanine residues in certain nucleotide sequences in liver cell DNA to form bulky adducts (see Chapter 2). In rodents, the main detoxification mechanism for aflatoxin

B_1-8-9-epoxide is through conjugation with glutathione catalyzed by GSTs. In these animals, the expression of GSTs can be induced by a synthetic dithiolethione derivative, *oltipraz*, previously used in humans as an anti-schistosomal agent. Oltipraz treatment was also observed to protect these animals from the hepatocarcinogenic effects of aflatoxin B_1. Based on these animal studies, oltipraz is currently being considered for human use to protect against aflatoxin B_1-induced hepatic cancer. Chemointervention is also possible at the dietary level. For instance, it will be recalled from Chapter 5 that many dietary sources contain plant phenolic antioxidants, and from Chapter 8 that allyl sulphide and diallyl sulphide, present in garlic and onions, can inhibit cytochrome P450s and activate GSTs.

In plants, GSTs play roles in the responses to auxin and during the normal metabolism of plant secondary products such as anthocyanins and cinnamic acid (see also Chapter 5). In addition, they are involved in the detoxification of a wide variety of xenobiotic compounds including herbicides. Besides their induction by auxins, many GSTs are induced by heavy metals, pathogen attack wounding, ethylene and ozone. It has been suggested that a common effect of all these processes is the generation of reactive oxygen species produced during oxidative stress, and that GSTs protect cellular components from oxidative damage. Plant GST promoters, unlike those from animal sources, do not contain functional XREs or EpREs. On the other hand, several contain *ocs* (octopine synthase) elements. These 20-base pair elements were first identified in promoters from plant pathogens such as cauliflower mosaic virus and *Agrobacterium tumifaciens*, and are activated by wounding. There is, however, some similarity between *ocs* elements and EpREs. Ocs elements contain the tandem core sequence, *ACGT*, and EpREs the tandem AP-1 sites. Some plant GST genes contain promoters that are more selectively induced. For instance, some contain domains also found in auxin regulatory elements and others contain ethylene-responsive elements.

9.8.2 MULTIDRUG RESISTANCE TRANSPORTER

As mentioned in Chapter 5, the mammalian *multidrug resistance transporter* is an important ATP-dependent transporter. Besides its normal cellular role in translocation of cellular lipids, it can function as an ATP-dependent efflux pump for many structurally diverse cytotoxic compounds.

Although the constitutive levels of expression of genes encoding the multidrug resistance transporter vary in different human and rodent tissues, studies in rat liver indicate that expression of the multidrug resistance gene (*mdr*) is modulated in response to toxic insults such as carcinogens. Additionally, *mdr* expression is dramatically elevated after exposure of rodent cells to a variety of chemotherapeutic agents that are known to be substrates for the multidrug transporter. While the effects of such chemotherapeutic drugs are not so clear-cut in human cells, the human *MDR1* gene can be activated by heat shock, arsenite and cadmium. Comparison of the promoters for the human *MDR1* and the rodent *mdr* genes has revealed a number of similarities and differences which may reflect the way in which these genes are regulated. Both types of gene have binding sites for the transcription factors AP-1 and Spi-1, but the human gene lacks a TATA-box. The regulatory differences include response to chemotherapy (the gene is activated in rodents but not in humans) and heat shock (the gene is activated in humans but not in rodents).

Multidrug resistance transporter genes are also found in bacteria, and a comparison of such genes in pathogenic and non-pathogenic bacteria has revealed very similar numbers. Such observations suggest that these transporters are not some recent evolutionary outcome in pathogens following antimicrobial chemotherapy.

9.8.3 MIMICS AND INAPPROPRIATELY REGULATED GENE EXPRESSION

In animals, hormones are produced and released into the bloodstream from endocrine glands as well as from testes and ovaries. Each hormone binds to a specific receptor associated with its target cells. These receptors often act as the initial switch-like component of signal transduction systems, which in certain cases can lead to the activation of specific genes encoding proteins, or enzymes, required for specific developmental pathways, or specialized cellular metabolic activity. For example, Figure 9.15 illustrates the binding of an oestrogenic hormone to a specific intracellular receptor. This receptor–hormone complex then migrates to the nucleus, where it binds to hormone response sequence elements in the upstream promoter regions of hormone-responsive genes to activate their transcription.

The genes for early sexual development, for example, are triggered by masculinizing steroid hormones (androgens, e.g. testosterone) and feminizing steroid hormones (oestrogens, e.g. oestradiol). These sex hormones are also involved in regulating genes controlling reproductive functions in adult life, such as the menstrual cycle and sperm production.

It now appears that a wide range of natural and synthetic chemicals share the ability to act like sex hormones, or can affect the

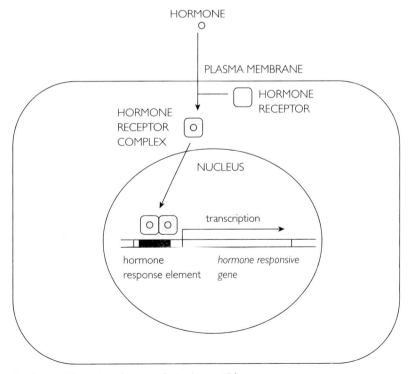

FIGURE 9.15 *A schematic illustration of gene regulation by steroid hormones*

TABLE 9.4 *Examples of chemical compounds with hormone disrupter activities*

Chemical compound	Comment
DDT	Organochlorine pesticides
Methoxychlor	
Chlordecone	
Atrazine	
Furans	Combustion and waste by-products
Dioxins	
Alkylphenol polyethoxylates	Surfactants (used in pesticides, paints, and cleaning products, and as processing aids in paper and textile production)
Bisphenol A	Breakdown product of polycarbonates, used in resins for can linings and white dental fillings
Nonylphenol	Softener for plastics
Polychlorinated biphenyls	Electrical insulators
Phthalate esters	Plasticizers found in food packaging plastics
Phytoestrogens (natural plant-derived oestrogen mimics)	Found in clover, parsley, sage, garlic, wheat, oats, rye, barley, hops, rice, cabbage, soybean, potatoes, carrots, beans, apples, cherries, plums, pomegranates, coffee, whisky
Vinclozolin	Fungicide

way in which natural sex hormones work. These are variously referred to as endocrine modulators or disruptors, xenoestrogens, environmental oestrogens or hormone mimics. In the case of oestrogenic compounds, the ability to bind to natural oestrogen receptors appears to be a relatively common phenomenon. Once bound, the mimics may act as agonists or antagonists, jamming the 'switch' permanently on or off. This can result in inappropriate expression, or lack of expression, of genes vital to a reproductive or developmental function. Some other chemicals are antiandrogenic and block androgen receptors; others interfere with the synthesis or metabolism of hormones or with their translocation within organisms. Table 9.4 lists some natural and synthetic compounds with *hormone disrupter activities* in animal systems.

In some animal experiments it has been demonstrated that exposure to some of these chemicals can alter sexual function and reproductive organs, or affect sexual development and function in the offspring of exposed pregnant mothers. For instance, 2,3,7,8-tetrachloro-*p*-dioxin (TCCD), which acts as an anti-oestrogen, harms sexual behaviour, sperm production and reproductive capacity in male rats. The fungicide vinclozolin, which behaves as an anti-androgen, damages reproductive organs in male offspring when given to pregnant rats. Certain phthalates reduce testosterone levels and shrink testes of male rats, and clover, rich in phytoestrogens, reduces sperm counts in rams. Additionally, some PCBs and dioxins, as well as thiocarbamide- and sulphonamide-based pesticides, can act as thyroid hormone disrupters. On balance, however, it should be pointed out that in many animal studies, doses of such disrupters far in excess of those found in food or the environment have been used. Moreover, the amount of natural oestrogens to which humans have been exposed are estimated to be many milliion times greater than the synthetic mimics. However, it should be emphasized that a reliable estimate of the total human exposure to natural and synthetic endocrine disrupters is not yet available. Nevertheless, in terms of direct effects of hormone mimics in humans, there is clear evidence of higher rates of reproductive abnormalities and dysfunction in the male and female children of mothers who took the drug diethylstilbestrol, previously used to prevent miscarriages.

A recent debate is whether there is a possible link between environmental chemicals that mimic or disrupt the action of oestrogens and breast cancer, which has increased considerably in the past 50 years. This is difficult to resolve due to the complexity of the disease. As in other malignancies, breast cancer is likely to arise when a cell escapes the normal restraints on replication and multiplies in an uncontrolled fashion. This requires the progressive accumulation of mutations in genes that control cell division and the maintenance of DNA. A further step is the specific 'promotion' of the growth of such abnormal cells and, in the case of breast cancer, oestrogenic hormones appear to play a central role in this latter process. A notable risk factor for breast cancer is sustained long-term exposure to oestradiol, for instance in those with early onset of menstruation or late entry into menopause, or never having had or breast-fed a child. If too much natural oestrogen is hazardous, it has been reasoned that prolonged exposure to hormone mimics or disrupters may also be risky. While hormone mimics or disrupters may enter the body in minute amounts they can accumulate to quite high levels, for instance in fatty tissues. Natural oestrogens, in contrast, are swiftly metabolized through normal pathways.

Environmental hormone disrupters are most likely to be ingested in dietary animal fat. Fatty foods from animals at the top of the food chain will contain the highest levels of disrupters, due to their progressive accumulation in the fatty tissues. Many patients with hormone-responsive breast cancers have been observed to have high serum levels of DDE (a primary metabolite of DDT) and polychlorinated biphenyls. Furthermore, those with higher cancer risk appear to have the highest serum levels of DDE. Research has indicated that oestradiol can be metabolised to 16-α-hydroxyoestrone and to 2-hydroxyoestrone (Figure 9.16). However, while the latter metabolite is protective, the former promotes the development of breast cancer. This has led to the suggestion that some hormone disrupters might encourage the growth of breast cancers by increasing the levels of 16-α-hydroxyoestrone in breast tissue.

An alternative view of the role of oestrogens in cancer has recently emerged. Rather than functioning as a promoter of breast cancer cell growth, cell culture experiments have shown that oestrogen metabolites can bind to DNA itself and thus cause gene damage. Thus, although hard data are lacking, oestrogens may well be involved in the initiating events as well as the promotion of carcinogenesis.

9.9 *Nutrient Stress*

9.9.1 RESPONSES TO NUTRIENT CHANGE IN BACTERIA

A considerable number of bacterial proteins are regulated by the presence, or absence, of a particular nutrient. Bacteria are generally noted for their ability to adjust extremely rapidly to nutrient changes, and they do this by regulating the production of proteins required for the breakdown, or synthesis, of particular compounds. There are many well-studied examples. For instance, the presence of lactose in the intestine causes the *E. coli* there

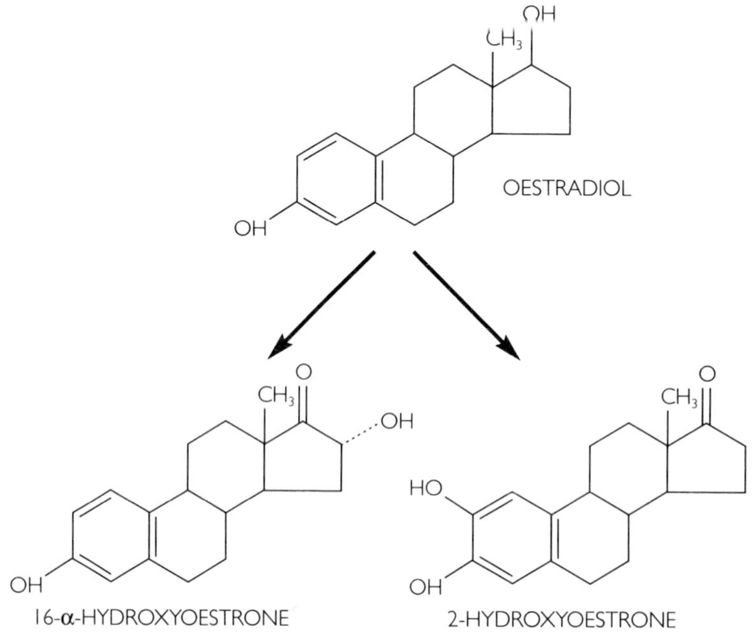

FIGURE 9.16 *Products of oestradiol metabolism*

to produce the enzyme β-galactosidase which cleaves the lactose to yield galactose and glucose. β-galactosidase is encoded by one of the genes, *lac z*, of the *lac operon*. The induction of this operon by lactose and the involvement of the catabolite activator protein and cyclic AMP, as well as the negative regulation of the operon by a repressor protein, has been described in Chapter 7. Another well-studied bacterial operon is the repressible *trp operon* which comprises the genes responsible for the synthesis of the amino acid tryptophan. The amount of tryptophan in the cell is regulated by blocking the transcription of the *trp* operon when there is sufficient tryptophan available, and the mechanism involving an aporepressor which is converted to an active repressor when complexed to tryptophan is also detailed in Chapter 7.

Faced with acute nutrient stress, certain bacteria react by forming spores; indeed, bacterial endospores have been described as the ultimate survivors. Vegetative cells of bacteria that form endospores, such as *Bacillus* and *Clostridium*, will grow and divide by binary fission until they become nutritionally deprived, at which time they may enter *sporulation*. The process is complex and involves many interacting regulatory pathways, transducing signals from many environmental and physiological stimuli. During sporulation, over 200 genes are differentially expressed under both temporal and spatial control. In *Bacillus subtilis* the initiation of sporulation is controlled by a multicomponent phosphorylation which serves to integrate signals from a variety of internal and external stimuli, and to channel this information to a single effector protein, the transcription factor *Spo0A*. Accumulation of the active phosphorylated form of Spo0A leads to the induction of a large number of genes. Among the the new proteins made are the two novel sigma-factors, σ^F and σ^E, which interact with RNA polymerase to initiate up-regulation of genes involved in the creation of the endospore, with its many

TABLE 9.5 *The degradative plasmids of Pseudomonads*

Plasmid	Function	Number of metabolic steps	Size (MDa)
CAM	Camphor catabolism	15–20	165
NAH	Naphthalene catabolism	>11	49
OCT	Octane catabolism	3	?
SAL	Salicylate catabolism	8	51
TOL	Toluene catabolism	>11	77
XYL	Xylene catabolism	>11	?

N.B. NAH, SAL and TOL all encode metabolic pathways leading to catechol.

unique structural features that confer its resistance characteristics.

In addition to gaining a clearer understanding of the role of novel sigma-factors in sporulation-gene regulation triggered by nutrient depletion, another recent advance has been the appreciation that bacterial cells that have stopped growing (*stationary-phase cells*) are more resistant to a variety of stresses such as heat, low pH and low water activity. Bacteria can enter the stationary phase for a number of reasons, but an important cause is nutrient depletion. In enteric bacteria the stationary-phase gene regulon is controlled by the RpoS protein, which codes for another novel sigma factor (σ^{38}) that substitutes in the RNA polymerase for the normal 'housekeeping' sigma factor (σ^{70}, the product of the *rpoD* gene) to bring about the specific up-regulation of transcription of a set of stationary-phase genes.

There are other situations in which the up-regulation of bacterial genes involves the substitution in RNA polymerase of the normal sigma-factor (σ^{70}) by another. One example, described in section 9.1, occurs in the case of the transcription of the *E. coli* heat-shock-inducible genes, which involves substitution of σ^{70} with σ^{32}, the product of the heat-inducible *rpoH* gene. Another example occurs in the up-regulation of transcription of bacterial genes for nitrogen metabolism. The *glnA* gene for *glutamine synthetase* requires RNA polymerase with σ^{54}, instead of σ^{70}, together with a specific activator protein (*nitrogen regulatory protein C*).

Another example of the nutritional versatility of bacteria is the ability of Pseudomonads (soil bacteria) to utilize a very wide range of synthetic and natural organic compounds as carbon and energy sources. The observation that some strains can use over 100 different chemical compounds made it unlikely that a bacterial genome alone could encode this number of catabolic functions. In fact, many of the compounds appear to be catabolized by plasmid-encoded functions (see Table 9.5).

In soil, many groups of bacteria, each carrying a particular type of plasmid that is transmissible to other neighbouring bacteria, would allow exploitation of a wide range of nutritional possibilities. The transfer of plasmids, or 'gene sharing', between strains may play an important role in the response of soil microflora to herbicides and other potentially toxic synthetic chemicals.

9.9.2 NURIENT RESPONSES IN ANIMAL AND PLANT CELLS

In comparison to bacteria, the responses to nutritional change and adversity have been less well studied in higher eukaryotes. However, as in microorganisms, sugar-sensitive genes are part of an ancient system of cellular adjustment to critical nutrient availability. For example, in the face of glucose deprivation, many mammalian cells are observed to upregulate the expression of the so-called *glucose regulated proteins (Grps)*, and, at the same time, to down-regulate the expression of heat-shock proteins (Hsps). Conversely, when glucose is restored, the synthesis of Grps is reduced and Hsp production restored. There is extensive amino acid homology (50%) between Hsp70s and the Grps of 75 kDa and 85 kDa. Moreover, Hsp90 is also extensively homologous to the Grp of 100 kDa. In fact, the Grps have many characteristics in common with Hsps. They all bind ATP and appear to function like chaperones. Grp75 is located in mitochondria, where it may serve in the assembly of mitochondrial proteins. Grp80 is believed to play a role in the folding of proteins as they pass through the endoplasmic reticulum, and Grp100 may play similar roles in the Golgi and plasma membrane.

Dietary sugars such as glucose and galactose are transported from the lumen of the mammalian intestine into enterocytes by the brush border *Na$^+$/D-glucose cotransporter* protein. In most species, the expression of the gene for this vital transporter is stimulated through a signal transduction pathway, initiated when the luminal sugar is perceived by a sugar sensor, located on the external face of the cells in the crypt region of the intestine. Positive gene responses to the presence of glucose are also observed with regard to a number of glycolytic and lipolytic genes in liver. *Glucose responsive elements (GIREs)* are found in the upstream promoter regions of the genes encoding enzymes, such as the *L-type pyruvate kinase* and *fatty acid synthase* as well as the hormone *insulin*. GIREs comprise two imperfect copies of the motif *CACGTG* separated by five base-pairs, to which 'upstream stimulatory factors' bind in order to elicit glucose responsiveness.

Plant gene responses to altered carbohydrate status are also varied, in as much as some are induced, some repressed and others only marginally affected. Sugar-regulated gene expression in plants provides a control of resource distribution between tissues and organs. Depletion of carbohydrates up-regulates genes for photosynthesis, remobilization and export, while down-regulating those for storage and utilization. Opposite effects accompany abundant sugar levels.

Homologues of the mammalian glucose-regulated proteins have been observed in plants as well as DNA-binding proteins, whose efficiency reflects cellular sugar status. At present it is hypothesized that that the rate of phosphorylation of hexoses by *hexokinase* is an important means of 'sensing' sugar levels. In yeast, hexokinase has a dual role, serving also as a protein kinase sensitive to the flux of sugars entering metabolism. It is proposed that the concentration of the complex between hexokinase and its product is directly involved in signalling the extent of carbon flux through the relevant metabolic pathways. It may be that the transfer of sugars across the plasma membrane is critical for sugar sensing and this may be closely coupled to hexokinase action. Other possible means of sensing sugar levels in plants, as in yeast and mammalian cells, relate to the relative concentrations of key metabolites such as acetate.

9.10 *Pathogen and Wounding Stress*

Although the environmental stresses discussed in this chapter have been abiotic in nature, brief mention also should be made of gene responses to infection by environmental pathogens. Cells within animals and plants are continually exposed to pathogen attack. To combat the threat of pathogenic microbes and viruses, vertebrates have developed highly complex mechanisms involving the interactions of many different cell types of the immune system to generate specific antibodies and cytotoxic cells. For a full description of such multicellular systems the reader is referred to standard immunology texts. In contrast to animal cells, each plant cell is capable of defending itself by a combination of constitutive and inducible defences. During the course of their coevolution, plants and pathogens have developed complex interrelationships. Pathogens have evolved an array of offensive mechanisms to parasitize plants and, in turn, plants have deployed a considerable array of mechanisms to achieve resistance.

Resistance to *plant pathogens* is often associated with a hypersensitive response involving localized induced cell death in the host plant at the site of infection, which is believed to limit pathogen growth. Following mutual recognition processes between the pathogenic organism and the target plant cell, the pathogen generates an array of metabolites including endopolygalacturonases that contribute to the release of pectic oligomers, or *endogenous elicitors*, which act as signal molecules. Binding of these elicitors to specific plant cell receptors, coded for by *plant R (resistance) genes*, causes membrane depolarization, which leads in turn to the activation of intracellular transduction systems to relay the signal to the genes of the plant cell nucleus. A recently isolated resistance gene is *Xa21* from rice, which confers resistance to the bacterium *Xanthomonas oryzae* pv. *Oryzae(Xoo)*, which causes blight in rice plants. The gene encodes the *protein Xa21*, which, as a protein kinase-like homodimer, spans the rice cell membrane to act as receptor for elicitor molecules specifically emanating from *Xoo*, and to activate cell defence mechanisms.

A commonly observed effect of elicitor binding to receptors appears to be the triggering of superoxide, and hydrogen peroxide generation involving a membrane-bound NADPH oxidase (see also Chapter 4). It may be that these active oxygen species play a role as intracellular second messengers in the processes of elicitor-induced plant gene activation, perhaps by controlling the redox status of key transcription factors. Other endogenous diffusable molecules that may function as messengers include *jasmonic acid, salicylic acid* and *ethylene*.

The plant genes that are induced following elicitor binding include those encoding structural compounds, enzymes of secondary metabolism, *pathogenesis-related proteins* and *protease inhibitors*. Primarily these are designed to protect the host tissues from colonization by the pathogen. The formation of structural molecules such as lignins, callose and hydroxyproline-rich glycoproteins reinforce the mechanical properties of the plant and limit the progress of the pathogen. Among the 'pathogenesis-related' proteins are *chitinases* and *β-1,3-glucanases* which can cause fungal wall hydrolysis with the release of chitin and β-1,3-glucan oligomers. These latter products are sometimes referred to as *exogenous elicitors* and can bind to specific plant membrane receptors to trigger events similar to those initiated by the endogenous elicitors. Protease inhibitor proteins are important because they function as an additional level of protection against not only damaging pathogens, but also potential predators. They achieve this by inhibiting many of the proteases that form a an essential part of the digestive system of most of the pathogens, pests, herbivores and humans that attack and eat plants.

Wounding of plants also elicits a specific stress response. Such injury might be the outcome of physical injury from chewing insects, and an initial event is the cross-linking of a specific protein within the cell-wall protein. This wound-induced cross-linking is similar to the formation of a blood clot in the circulatory system in animals, and is followed by the increased synthesis of the polysaccharide *callose*, by a plasma membrane enzyme system, which also serves to repair the physical damage. The *wounding response* also causes the increased expression of genes for protease inhibitors to limit any further predator damage as described above. Despite these advances in our appreciation of these responses to wounding, the means whereby perception of the wound leads to specific gene activation remains uncertain. Wounding will lead to the loss of cellular integrity and abnormal mixing of hydrolytic enzymes of the vacuoles with components of the cell wall, resulting in the hydrolysis of cell wall polysaccharides and the release of oligomeric fragments. These fragments may be recognized by plasma membrane components (possibly receptors) and elicit a change in the transmembrane flow of ions which may be instrumental in initiating the wounding response. Overall, however, it appears that some type of mobile signal generated by the wound is transmitted around the rest of the plant to elicit a systemic response. Contenders for such a mobile signal include *abscisic acid* and *jasmonic acid*.

9.11 *Literature Sources and Additional Reading*

9.11.1 HIGH-TEMPERATURE STRESS

BURDON, R.H., 1993, Heat shock proteins in relation to medicine, *Molecular Aspects of Medicine*, 14, 83–165.

BURDON, R.H., 1993, Stress proteins in plants, *Botanical Journal of Scotland*, 46, 463–75.

CONWAY DE MACARIO, E. and MACARIO, A.J.L., 1994, Heat shock response in *Archaea, Trends in Biotechnology*, 12, 512–18.

HARTL, F.U., 1996, Molecular chaperones in cellular protein folding, *Nature*, 381, 571–80.

MIERNYK, J.A., 1997, The 70 kDa stress-related proteins as molecular chaperones, *Trends in Plant Science*, 2, 180–84.

MORIMOTO, R.I., KROEGER, P.E. and COTTO, J.J., 1996, The transcriptional regulation of heat shock genes: a plethora of heat shock factors and regulatory conditions, in Fiege, U., Morimoto, R.I., Yahara, I. and Polla, B. (Eds) *Stress-inducible Cellular Responses*, pp. 139–63, Basel: Birkhauser Verlag.

MORIMOTO, R.I., TISSIERES, A. and GEORGOPOULOS, C., 1990, *Stress Proteins in Biology and Medicine*, Cold Spring Harbor, New York: Cold Spring Harbor Laboratory Press.

MORIMOTO, R.I., TISSIERES, A. and GEORGOPOULOS, C., 1994, *The Biology of Heat Shock Proteins and Molecular Chaperones*, Cold Spring Harbor, New York: Cold Spring Harbor Laboratory Press.

NAGAO, R.T., KIMPEL, J.A. and KEY, J.L., 1990, Molecular and cellular biology of the heat shock response, in Scandalios, J.G. (Ed.) *Genomic Responses to Environmental Stress*, pp. 235–74, San Diego, CA: Academic Press.

VIERLING, E., 1991, The roles of heat shock proteins in plants, *Annual Review of Plant Physiology and Plant Molecular Biology*, 42, 579–620.

VOELLMY, R., 1996, Sensing stress and responding to stress, in Fiege, U., Morimoto, R.I., Yahara, I. and Polla, B. (Eds) *Stress-inducible Cellular Responses*, pp. 121–137, Basel: Birkhauser Verlag.

WATERS, E.R., LEE, G.J. and VIERLING, E., 1996, Evolution, structure and function of the small heat shock proteins in plants, *Journal of Experimental Botany*, 47, 325–38.

WELCH, W.J., 1993, How cells respond to stress, *Scientific American*, May, 34–41.

YURA, T., NAKAHIGASHI, K. and KANEMORI, M., 1996, Transcriptional regulation of stress-inducible genes in procaryotes, in Fiege, U., Morimoto, R.I., Yahara, I. and Polla, B. (Eds) *Stress-inducible Cellular Responses*, pp. 165–81, Basel: Birkhauser Verlag.

9.11.2 LOW-TEMPERATURE STRESS

CHANDLER, P.M. and ROBERTSON, M., 1994, Gene expression regulated by abscisic acid and its relation to stress tolerance, *Annual Review of Plant Physiology and Plant Molecular Biology*, 45, 113–41.

CRAWFORD, R.M.M., 1989, *Studies in Plant Survival, Ecological Case Histories of Plant Adaptation to Adversity*, Oxford: Blackwell Scientific Publications.

FLEURY, C., BOUILLAUD, F. and RICQUIER, D., 1997, A plant cold-induced uncoupling protein, *Nature*, 389, 135–36.

GUY, C.L., 1990, Cold acclimation and freezing stress tolerance: role of protein metabolism, *Annual Review of Plant Physiology and Plant Molecular Biology*, 41, 187–223.

HARWOOD, J., 1991, Strategies for coping with low environmental temperatures, *Trends in Biochemical Sciences*, 16, 126–27.

HUGHES, M.A. and DUNN, M.A., 1996, The molecular biology of plant acclimation to low temperature, *Journal of Experimental Botany*, 47, 291–305.

JIA, Z., DeLUCA, C.I., CHAO, H. and DAVIES, P.L., 1996, Structural basis for the binding of a globular antifreeze protein to ice, *Nature*, 384, 285–88.

MURATA, N. and LOS, D.A., 1997, Membrane fluidity and temperature perception. *Plant Physiology*, 115, 875–79.

NIHIDA, I. and MURATA, N., 1996, Chilling sensitivity in plants and cyanobacteria; the crucial contribution of membrane lipids, *Annual Review of Plant Physiology and Plant Molecular Biology*, 47, 541–68.

SCHINDELIN, H., MARAHIEL, M.A. and HEINEMANN, U., 1993, Universal nucleic acid-binding domain revealed by crystal structure of the *B. subtilis* major cold-shock protein, *Nature*, 364, 164–68.

SHMIDT-NIELSEN, K., 1983, *Animal Physiology: Adaptation and Environment*, 3rd edn, pp. 225–48, Cambridge: Cambridge University Press.

THOMASHOW, M.F., 1998, Role of cold-responsive genes in plant freezing tolerance, *Plant Physiology*, 118, 1–8.

9.11.3 WATER STRESS

CELLIER, F., CONEJERO, G., BREITLER, J.-C. and CASSE, F., 1998, Molecular and physiological responses to water deficit in drought-tolerant and drought-sensitive lines of sunflower. Accumulation of dehydrin transcripts correlates with tolerance, *Plant Physiology*, 116, 319–28.

CHANDLER, P.M. and ROBERTSON, M., 1994, Gene expression regulated by abscisic acid and its relation to stress tolerance, *Annual Review of Plant Physiology and Plant Molecular Biology*, 45, 113–41.

DORMAN, C.J., 1996, Flexible response: DNA supercoiling, transcription and bacterial adaptation to environmental stress, *Trends in Microbiology*, 4, 214–16.

INGRAM, J. and BARTELS, D., 1996, The molecular basis of dehydration tolerance in plants, *Annual Review of Plant Physiology and Plant Molecular Biology*, 47, 377–403.

ISHITANI, M., XIONG, L., STEVENSON, B. and ZHU, J.-K., 1997, Genetic analysis of osmotic and cold stress transduction in *Arabidopsis*: interactions and convergence of abscisic acid dependent and abscisic acid independent pathways, *Plant Cell*, 9, 1935–49.

LEPRINCE, O., HENDRY, G.A.F. and MCKERSIE, B.D., 1993, The mechanisms of desiccation tolerance in developing seeds, *Seed Science Research*, 3, 231–46.

MCCUE, K.F. and HANSON, A.D., 1990, Drought and salt tolerance: towards understanding and application, *Trends in Biotechnology*, 8, 358–62.

RUSSELL, B.L., RATHINASABAPATHI, B. and HANSON, A.D., 1998, Osmotic stress induces expression of choline monooxygenase in sugar beet and amaranthus, *Plant Physiology*, 116, 859–65.

SHINOZAKI, K. and YAMAGUCHI-SHINOZAKI, K., 1997, Gene expression and signal transduction in water-stress response, *Plant Physiology*, 115, 327–34.

STOOP, J., WILLIAMSON, J.D. and MASON PHARR, D., 1996, Mannitol metabolism in plants: a method of coping with stress, *Trends in Plant Science*, 1, 139–44.

TROSSAT, C., RATHINASABAPATHI, B., WERETILNYK, E.A., SHEN, T.-L., HUANG, Z.-H., GAGE, D.A. and HANSON, A.D., 1998, Salinity promotes accumulation of 3-dimethylsulphonioproprionate and its precursor S-methylmethionine in chloroplasts, *Plant Physiology*, 116, 165–71.

YEO, A., 1998, Molecular biology of salt tolerance in the context of whole plant physiology, *Journal of Experimental Botany*, 49, 915–24.

9.11.4 TOXIC METAL STRESS

FISCHER, E.H. and DAVIE, E.W., 1998, Recent excitement regarding metallothionein, *Proceedings of the National Academy of Sciences USA*, **95**, 3333–34.

MERGEAY, M., 1991, Towards an understanding of the genetics of bacterial metal resistance, *Trends in Biotechnology*, **9**, 17–24.

PENA, M.O., KOCH, K.A. and THIELE, D.J., 1998, Dynamic regulation of copper uptake and detoxification in *Saccharomyces cerevisiae*, *Molecular and Cellular Biology*, **18**, 2514–23.

ROBINSON, N.J., TOMMEY, A.M., KUSKE, C. and JACKSON, P.J., 1993, Plant metallothioneins, *Biochemical Journal*, **295**, 1–10.

SUMMERS, A.O., 1985, Bacterial resistance to toxic elements, *Trends in Biotechnology*, **3**, 122–25.

THIELE, D., 1992, Metal-regulated transcription in eukaryotes, *Nucleic Acids Research*, **20**, 1183–91.

ZHU, Z. and THIELE, D.J., 1996, Toxic metal-responsive gene transcription, in Fiege, U., Morimoto, R.I., Yahara, I. and Polla, B. (Eds) *Stress-inducible Cellular Responses*, pp. 307–20, Basel: Birkhauser Verlag.

9.11.5 OXIDATIVE STRESS

ASADA, K., 1997, The role of ascorbate peroxidase and monohydroascorbate reductases in H_2O_2 scavenging in plants, in Scandalios, J.G. (Ed.) *Oxidative Stress and the Molecular Biology of Antioxidant Defenses*, pp. 715–35, Cold Spring Harbor, NY: Cold Spring Harbor Laboratory Press.

BOWLER, C., VAN MONTAGU, M. and INZE, D., 1992, Superoxide dismutase and stress tolerance, *Annual Review of Plant Physiology and Plant Molecular Biology*, **43**, 83–116.

BURDON, R.H., 1997, Oxyradicals as signal inducers, in Forman, H.J. and Cadenas, E. (Eds) *Oxidative Stress and Signal Transduction*, pp. 289–322, New York: Chapman and Hall.

BURDON, R.H., GILL, V. and RICE-EVANS, C., 1990, Active oxygen species and heat shock protein induction, in Shlesinger, M.J., Santoro, M.G. and Garaci, E. (Eds) *Stress Proteins*, pp. 16–25, Berlin: Springer Verlag.

DAVIES, J.M.S., LOWRY, C.V. and DAVIES, K.J.A., 1995, Transient adaptation to oxidative stress in yeast, *Archives of Biochemistry and Biophysics*, **317**, 1–6.

DAVIES, K.J.A., and LI, S.W., 1994, Bacterial gene expression during oxidative stress, in Nohl, H., Esterbauer, H. and Rice-Evans, C. (Eds), *Free Radicals in the Environment, Medicine and Toxicology*, pp. 563–578, London: Richelieu Press.

FLOHE, L., WINGENDER, E. and BRIGELIUS-FLOHE, R., 1997, in Forman, H.J. and Cadenas, E. (Eds) *Oxidative Stress and Signal Transduction*, pp. 415–40, New York: Chapman and Hall.

HILDAGO, E., DING, H. and DEMPLE, B., 1997, Redox signal transduction via iron–sulphur clusters in the SoxR transcription activator, *Trends in Biochemical Sciences*, 22, 207–210.

JAMIESON, D.J. and STORZ, G., 1997, Transcriptional regulators of oxidative stress responses, in Scandalios, J.G. (Ed.) *Oxidative Stress and the Molecular Biology of Antioxidant Defenses*, pp. 91–115, Cold Spring Harbor, NY: Cold Spring Harbor Laboratory Press.

JANSSEN, Y.M.W., TIMBLIN, C.R., ZANELLA, C.L., JIMENEZ, J.A. and MOSSMAN, B.T., 1997, Induction of gene expression by environmental oxidants associated with inflammation, fibrogenesis and carcinogenesis, in Forman, H.J. and Cadenas, E. (Eds) *Oxidative Stress and Signal Transduction*, pp. 387–414, New York: Chapman and Hall.

MELHORN, H., 1990, Ethylene-promoted ascorbate peroxidase activity protects plants against hydrogen peroxide, ozone and paraquat, *Plant, Cell and Environment*, 13, 971–76.

MULLINEAUX, P.M. and CREISSEN, G.P., 1997, Glutathione reductases: regulation and role in oxidative stress, in Scandalios, J.G. (Ed.) *Oxidative Stress and the Molecular Biology of Antioxidant Defenses*, pp. 667–713, Cold Spring Harbor, NY: Cold Spring Harbor Laboratory Press.

SCANDALIOS, J.G., 1997, Molecular genetics of superoxide dismutases in plants, in Scandalios, J.G. (Ed.) *Oxidative Stress and the Molecular Biology of Antioxidant Defenses*, pp. 527–68, Cold Spring Harbor, NY: Cold Spring Harbor Laboratory Press.

SCANDALIOS, J.G., GUAN, L. and POLIDOROS, A.N., 1997, in Scandalios, J.G. (Ed.) *Oxidative Stress and the Molecular Biology of Antioxidant Defenses*, pp. 343–406, Cold Spring Harbor, NY: Cold Spring Harbor Laboratory Press.

SCHULZE-OSTHOFF, K., BAUER, M., VOGT, M., WESSELBORG, S. and BAEUERLE, P.A., 1997, in Forman, H.J. and Cadenas, E. (Eds) *Oxidative Stress and Signal Transduction*, pp. 239–59, New York: Chapman and Hall.

SEN GUPTA, A., WEBB, R.P., HOLADAY, A.S. and ALLEN, R.D., 1993, Overexpression of superoxide dismutase protects plants from oxidative stress, *Plant Physiology*, 103, 1067–73.

STORZ, G. and POLLA, B.S., 1996, Transcriptional regulators of oxidative stress-inducible genes, in Fiege, U., Morimoto, R.I., Yahara, I. and Polla, B. (Eds) *Stress-inducible Cellular Responses*, pp. 239–54, Basel: Birkhauser Verlag.

SUN, Y. and OBERLEY, L.W., 1996, Redox regulation of transcriptional activators, *Free Radical Biology and Medicine*, **21**, 335–48.

WIESE, A.G., PACIFICI, R.E. and DAVIES, K.J.A., 1995, Transient adaptation to oxidative stress in mammalian cells, *Archives of Biochemistry and Biophysics*, **318**, 231–40.

ZHENG, M., ASLUND, F. and STORTZ, G., 1998, Activation of the OxyR transcription factor by reversible disulphide bond formation, *Science*, **279**, 1718–21.

9.11.6 HYPOXIC AND ANOXIC STRESS

AN, W.G., KANEKAL, M., SIMON, M.C., MALTEPE, E., BLAGOSKLONNY, M.K. and NECKERS, L.M., 1998, Stabilization of wild-type p53 by hypoxia-inducible factor 1α, *Nature*, **329**, 405–408.

CRAWFORD, R.M.M., 1989, *Studies in Plant Survival, Ecological Case Histories of Plant Adaptation to Adversity*, Oxford: Blackwell Scientific Publications.

CRAWFORD, R.M.M., 1993, Plant survival without oxygen, *Biologist*, **40**, 110–14.

DREW, M.C., 1997, Oxygen deficiency and root metabolism: injury and acclimation under hypoxia and anoxia, *Annual Review of Plant Physiology and Plant Molecular Biology*, **48**, 223–50.

GOLDBERG, M.A., DUNNING, S.P. and BUNN, H.F., 1988, Regulation of the erythropoetin gene. Evidence that the oxygen sensor is a haem protein, *Science*, **242**, 1412–15.

GUILLEMIN, K. and KRASNOW, M.A., 1997, The hypoxic response: huffing and HIFing, *Cell*, **89**, 9–12.

TACCHINI, L., SCHIFFONATI, L., PAPPALARDO, C., GATTI, S. and BERNELLI-ZAZZERA, A., 1993, Expression of HSP70, immediate-early response and haem oxygenase genes in ischaemic-reperfused rat liver, *Laboratory Investigation*, **68**, 465–71.

WANG, G.L. and SEMENZA, G.L., 1993, Characterization of hypoxia-inducible factor 1 and regulation of DNA binding by hypoxia, *Journal of Biological Chemistry*, **268**, 21513–18.

9.11.7 ULTRAVIOLET RADIATION STRESS

FROHNMEYER, H., BOWLER, C. and SCHAFER, E., 1997, Evidence for some signal transduction elements involved in UV-light-dependent responses in parsley protoplasts, *Journal of Experimental Botany*, 48, 739–50.

POLLE, A., 1997, Defense against photooxidative damage in plants, in Scandalios, J.G. (Ed.) *Oxidative Stress and the Molecular Biology of Antioxidant Defenses*, pp. 623–66, Cold Spring Harbor, NY: Cold Spring Harbor Laboratory Press.

RYTER, S.F. and TYRELL, R.M., 1997, The role of haem oxygenase-1 in the mammalian stress response: molecular aspects of regulation and function, in Forman, H.J. and Cadenas, E. (Eds) *Oxidative Stress and Signal Transduction*, pp. 343–86, New York: Chapman and Hall.

RYTER, S.W. and TYRELL, R.M., 1998, Singlet molecular oxygen ($^1O_2$): a possible effector of eukaryotic gene expression, *Free Radical Biology and Medicine*, 24, 1520–34.

SHINIGAWA, H., 1996, SOS response as an adaptive response to DNA damage in prokaryotes, in Fiege, U., Morimoto, R.I., Yahara, I. and Polla, B. (Eds) *Stress-inducible Cellular Responses*, pp. 221–35, Basel: Birkhauser Verlag.

TYRELL, R.M., 1996, UV activation of mammalian stress proteins, in Fiege, U., Morimoto, R.I., Yahara, I. and Polla, B. (Eds) *Stress-inducible Cellular Responses*, pp. 256–71, Basel: Birkhauser Verlag.

WELLBURN, A., 1994, *Air Pollution and Climate Change, the Biological Impact*, pp. 145–69, Harlow: Longman Scientific and Technical.

9.11.8 TOXIC CHEMICAL STRESS

BRZOZOWSLI, A.M., PIKE, A.C.W., DAUTER, Z., HUBBARD, R.E., BONN, T., ENGSTROM, O., OHMAN, L., GREENE, G.L., GUSTAFSSON, J.-A. and CARLQUIST, M., 1997, Molecular basis of agonism and antagonism in the oestrogen receptor, *Nature*, 389, 753–58.

BUETLER, T.M., BAMMLER, T.K., HAYES, J.D. and EATON, D.L., 1996, Oltipraz-mediated changes in aflatoxin B_1 transformation of in rat liver: implications for human chemointervention, *Cancer Research*, 56, 2306–13.

CHANG, C.-Y. and PUGA, A., 1998, Constitutive activation of the aromatic hydrocarbon receptor, *Molecular and Cellular Biology*, 18, 525–35.

CHIN, K.-V., PASTAN, I. and GOTTESMAN, M.M., 1993, Function and regulation of the human multidrug resistance gene, *Advances in Cancer Research*, **60**, 157–80.

COON, M.J., DING, X., PERNECKY, S.J. and VAZ, A.D.N., 1992, Cytochrome P450: progress and predictions, *FASEB Journal*, **6**, 669–73.

DANIEL, V., 1993, Glutathione S-transferases: gene structure and regulation of expression, *Critical Reviews in Biochemistry and Molecular Biology*, **28**, 173–207.

DAVIS, D.L., BRADLOW, H.L., WOLFF, M., WOODRUFF, T., HOEL, D.G. and ANTON-CULVER, H., 1993, Medical hypothesis: xenoestrogens as preventable causes of breast cancer, *Environmental Health Perspectives*, **101**, 372–377.

FAVREAU, L.V. and PICKETT, C.B., 1997, The antioxidant response element, in Forman, H.J. and Cadenas, E. (Eds) *Oxidative Stress and Signal Transduction*, pp. 272–288, New York: Chapman and Hall.

GONZALEZ, F.J. and NEBERT, D.W., 1990, Evolution of the P450 gene superfamily, *Trends in Genetics*, **6**, 182–86.

MARRS, K.A., 1996, The functions and regulation of glutathione S-transferases in plants, *Annual Review of Plant Physiology and Plant Molecular Biology*, **47**, 127–58.

PRESTERA, T. and TALALAY, P., 1995, Electrophile and antioxidant regulation of enzymes that detoxify carcinogens, *Proceedings of the National Academy of Sciences USA*, **92**, 8965–69.

PROCHASKA, H.J. and TALALAY, P., 1988, Regulatory mechanisms of monofunctional and bifunctional anticarcinogenic enzyme inducers in murine liver, *Cancer Research*, **48**, 4776–82.

SAIER, M.H., PAULSEN, I.T., SLIWINSKI, M.K., PAO, S.S., SKURRAY, R.A. and NIKAIDO, H., 1998, Evolutionary origin of multi-drug and drug-specific efflux pumps in bacteria, *FASEB Journal*, **12**, 265–74.

SERVICE, R.F., 1998, New role for oestrogen in cancer, *Science*, **279**, 1631–33.

SILVERMAN, J.A. and SCHRENK, D., 1997, Expression of the multidrug resistance genes in the liver, *FASEB Journal*, **11**, 308–13.

THOMAS, J.A., 1997, Phytoestrogens and hormonal modulation: a mini-review, *Environmental and Nutritional Interactions*, **1**, 5–12.

WU, L. and WHITLOCK JR, J.P., 1993, Mechanism of dioxin action: receptor–enhancer interactions in intact cells, *Nucleic Acids Research*, **21**, 119–25.

9.11.9 NUTRIENT STRESS

D'AMINICO, L., VALSASINA, B., DAMATINI, M.G., FABBRINI, M.S. and NITTI, G., 1992, Bean homologues of the mammalian glucose regulated proteins: induction by tunicamycin and interaction with newly synthesized seed storage proteins in the endoplasmic reticulum, *Plant Journal*, 2, 443–55.

ERRINGTON, J., 1993, *Bacillus subtilis* sporulation: regulation of gene expression and control of morphogenesis, *Microbiological Reviews*, 57, 1–33.

GANCEDO, J.M., 1992, Carbon catabolite repression in yeast, *European Journal of Biochemistry*, 206, 297–313.

KOCH, K.E., 1996, Carbohydrate-modulated gene expression in plants, *Annual Review of Plant Physiology and Plant Molecular Biology*, 47, 509–40.

PELHAM, H.R.B., 1986, Speculation on the function of the major heat shock and glucose regulated genes, *Cell*, 46, 959–61.

ROSE, M., ALBIG, W. and ENTIAN, K.D., 1991, Glucose repression in *Saccharomyces cerevisiae* is directly associated with hexose phosphorylation by hexokinases, *European Journal of Biochemistry*, 199, 511–18.

TRUMBLY, R.J., 1992, Glucose repression in the yeast *Saccharomyces cerevisiae*, *Molecular Microbiology*, 6, 15–21.

WELCH, W.J., 1990, The mammalian stress response: cell physiology and biochemistry of stress proteins, in Morimoto, R.I., Tissieres, A. and Georgopoulous, C. (Eds) *Stress Proteins in Biology and Medicine*, pp. 223–78, Cold Spring Harbor, NY: Cold Spring Harbor Laboratory Press.

9.11.10 PATHOGEN AND WOUNDING STRESS

BENHAMOU, N., 1996, Elicitor-induced plant defense pathways, *Trends in Plant Sciences*, 1, 233–40.

DRAPER, J., 1997, Salicylate, superoxide synthesis and cell suicide in plant defense, *Trends in Plant Science*, 2, 162–65.

DURNER, J., SHAH, J. and KLESSIG, D.F., 1997, Salicylic acid and disease resistance in plants, *Trends in Plant Science*, 2, 266–74.

FARMER, E.E. and RYAN, C.A., 1992, Octadecanoid-derived signals in plants, *Trends in Cell Biology*, 2, 236–41.

KEEN, N.T., 1992, The molecular biology of disease resistance, *Plant Molecular Biology*, **19**, 109–22.

STASKAWICZ, B.J., AUSUBEL, F.M., BAKER, B.J., ELLIS, J.G. and JONES, J.A., 1995, Molecular genetics of plant disease resistance, *Science*, **268**, 661–67.

CHAPTER

Stress Genes and Biotechnology

KEY POINTS

- Specific stress genes have been genetically engineered in stress-sensitive plants with the object of producing transgenic species more capable of withstanding particular environmental stresses.

- The measurement of specific stress gene activation can be used as a toxicological end-point, and thus as the basis of 'biosensors' of environmental stress.

- Stress-resistant proteins from extremophiles may be valuable as environmentally clean catalysts in industrial processes.

- Genes encoding xenobiotic degradation enzymes and toxic metal metabolism offer potential as a future route to bioremediation.

The previous chapter highlighted the large body of information, accumulated over the past two decades, that elucidates the ways in which organisms sense and respond to various environmental pollutants and stressors. Perhaps the most knowledge concerns the structure of an impressive number of stress genes and the mechanisms whereby these genes can be activated, as well as the properties of their products in relation to cell survival. Not surprisingly, potential biotechnological applications in agriculture, brewing, baking, medicine, pharmacology, and toxicology are beginning to emerge. For example, with the advent of genetic engineering, the possibility of introducing specific stress genes directly into stress-sensitive organisms offers a rapid and novel route to the development of stress-tolerant crops. Another potential application is the use of stress gene activation and stress proteins as biomarkers for estimating exposure, or damage, from environmental exposure to toxic chemicals and other stressors. Future biotechnological applications can also be envisaged for a range of proteins derived from extremophiles, a class of microorganisms that inhabit a range of adverse environments, for example high temperatures or extremes of pH. Stress-tolerant proteins from these organisms could be of considerable value as catalysts for environmentally clean industrial processes. Although yet to be fully explored, knowledge of toxic metal metabolizing genes and genes encoding components of xenobiotic degradation pathways offers potential as a future route to environmental bioremediation.

10.1 Genetic Engineering for Stress Tolerance

Despite the initial optimism and conceptual attractiveness of genetic engineering techniques to insert stress genes into stress-sensitive organisms and achieve stress tolerance, progress in this direction has been disappointing. Many environmental stresses, rather than activating the expression of single gene products, more often result in the activation of several genes. Moreover, several stress genes are activated by more than one environmental stressor. Although there is considerable knowledge regarding stress genes and their products, the achievement of stress-tolerance in many organisms often involves the complex interplay of many stress gene products and multiple gene activators within the context of intrinsic cell physiological processes. As a result, it is often still difficult to provide a coherent molecular explanation of acclimation responses. Notwithstanding these difficulties, some studies have shown that it is possible, in certain cases, to alter the stress tolerance of complex multicellular organisms by deliberate genetic intervention.

10.1.1 THERMAL TOLERANCE

In the fruit fly, *Drosophila melanogaster* (unlike *E. coli* and yeast), *Hsp70* appears to be the major protein involved in tolerance to extreme temperatures, and cultured *Drosophila* cells transformed with extra copies of the *hsp70* gene acquire thermotolerance at a faster rate than wild-type cells. Moreover, when *hsp70* is expressed from a heterologous promoter at normal temperatures (25°C), this is sufficient to provide tolerance to direct exposure to 42°C. At the organismal level, fly strains carrying 12 extra copies of the wild-type *hsp70* gene demonstrated a more rapid acquisition of thermotolerance.

In vertebrates, there is also evidence that transformation of hsp70 genes into cultured cells, of monkeys and rats, dramatically increases cellular thermotolerance. There are, however, no data at the organismal level.

10.1.2 CHILLING TOLERANCE

The sensitivity of plants to chilling is related to the degree of unsaturation of fatty acids in the phosphatidylglycerol of chloroplast membranes. Those plants with high proportions of cis-unsaturated fatty acids are resistant to chilling, whereas those with a small proportion are not. The proportion of unsaturated fatty acids in phosphatidylglycerol appears to be controlled by the chloroplast enzyme *glycerol-3-phosphate acetyltransferase*. By transformation of tobacco plants with genes for glycerol-3-phosphate acetyltransferase from squash, or *Arabidopsis*, the level of fatty acid saturation and the degree of chilling sensitivity can be manipulated in tobacco plants. While factors other than the extent of unsaturation of the fatty acids in phosphatidylglycerol may affect chilling sensitivity, these approaches demonstrate that this is, nevertheless, an important factor.

As detailed in the previous chapter (section 9.2), many plants show increased resistance to freezing after exposure to low, non-freezing, temperatures. This response is associated with the induction of a variety of low-temperature responsive (LTR) genes. In *Arabidopsis* this is mediated by the binding of a transcription factor, CBF1, to common upstream promoter sequences. A recent novel approach has been to engineer the overexpression of the gene for CBF1 in *Arabidopsis* to produce plants with increased freezing tolerance.

10.1.3 PATHOGEN RESISTANCE

Bacterial blight is a considerable hazard to rice growers. *Xanthomonas oryzae* pv. *Oryzae* spreads rapidly from plant to plant, and from field to field, in water droplets. Upon infection rice leaves develop lesions and quickly wilt. As a result of work aimed at transferring blight resistance from wild rice to the cultivated variety, using conventional breeding techniques, it emerged that resistance could conferred by a single gene of wild rice, namely *Xa21*. This gene encodes a protein *Xa21* which, in its homodimeric form, can span the rice cell membrane and act as receptor specific for elicitors produced from the infecting *Xanthomonas*. As mentioned in the previous chapter, the binding of pathogen elicitor molecules by such receptors triggers the activation of a cascade of plant genes required to provide protection from the pathogen. Transfer of the isolated *Xa21* gene into blight-sensitive rice strains resulted in pathogen-resistant plants. Moreover, this resistance was passed on to the next generation of plants through self-fertilization.

10.1.4 DEHYDRATION TOLERANCE

The accumulation of low molecular weight metabolites that act as osmoprotectants is a common adaptation to dry or saline conditions in many organisms (section 9.3). Transgenic tobacco plants containing the bacterial gene for *mannitol 1-phosphate dehydrogenase* synthesize and accumulate *mannitol* and have an increased tolerance to salt. Tobacco plants have also been engineered to carry bacterial *fructosyltransferase* genes. They accumulate the polyfructose molecule *fructan* and, under certain conditions of drought stress, show improved growth. However, the mechanism by which fructans confer such drought tolerance is unknown.

The genetic modification of the ability of plants to tolerate salt has also been approached by transferring to *Arabidopsis* a gene from *Arthrobacter globiformis* that encodes *choline oxidase*, the enzyme that catalyzes the oxidation of choline to *glycine betaine*. Such transgenic plants accumulate the osmoprotectant glycine betaine and are able to grow in 100mM NaCl, as well as showing some increased tolerance to cold stress.

10.1.5 OXIDATIVE STRESS TOLERANCE

The protective role of superoxide dismutases against oxidative stress in bacteria has been supported by studies involving a range of mutants. Transgenic experiments have demonstrated that plant superoxide dismutases can be effectively utilized to rescue superoxide dismutase deficient yeasts from oxidative stress. In tobacco plants themselves, overexpression of *Mn-superoxide dismutase* has been shown to provide protection from the oxidative stress induced by exposure of the transgenic plants to the herbicide paraquat, although moderate overexpression of *Cu,Zn-superoxide dismutase* was not protective.

10.1.6 HERBICIDE TOLERANCE

Although transgenic plants overexpressing superoxide dismutase activity have increased tolerance to the herbicide paraquat, they are not tolerant of herbicides whose mode of action does not involve the promotion of increased cellular generation of superoxide. Much commercial interest has recently focused on the development of transgenic plants tolerant to other herbicides. Although the approaches used do not actually involve the use of any of the stress genes described in the previous chapter, the outcomes have served to signal scientific and public concerns regarding the propagation and environmental release of genetically engineered organisms, which might in the future contain stress genes. In this context of concern, a particularly significant example has been the genetic engineering

of resistance to the herbicide glyphosate into commercially important crop species such as soybean and oilseed rape.

Unlike mammals, plants and most microbial species have the capacity to synthesize all the major amino acids. However, because mammals lack the biosynthetic pathways for phenylalanine, tryptophan, lysine, leucine, isoleucine, valine, threonine, and methionine, these preformed amino acids are essential components of the mammalian diet, together with arginine and histidine in the case of immature animals. For this reason, the biosynthetic pathways to these particular amino acids, in plants and microbes, are particularly attractive targets for toxicologically safe herbicides, and pesticides aimed at plant pathogens. In green plant tissues, erythrose-4-phosphate and phosphoenolpyruvate (PEP) are the products of the Calvin cycle and glycolysis respectively and are the starting materials for the seven-step *shikimate pathway* leading to the formation of precursors of aromatic amino acids. An important catalyst in this pathway is the enzyme *5-enolpyruvylshikimate-3-phosphate synthase (EPSP synthase)*, which promotes the reversible transfer of the enolpyruvate group from PEP to shikimate 3-phosphate in chloroplasts. EPSP synthases from bacteria and plants are monomeric monofunctional proteins of 40–50 kDa. Critically, the broad spectrum herbicide *glyphosate* acts as a specific and potent inhibitor of EPSP synthases, and in plants it has been established that this enzyme is the target site for its herbicidal action.

The gene for EPSP synthase in plants is located in the nucleus although its product is required within chloroplasts. Conventional gene transfer strategies for producing glyphosate-tolerant plants are to insert an EPSP gene from a plant or a tolerant bacterium into the plant cell nucleus, but with suitable genetic signal sequences to target the product to chloroplasts. The aim is to increase the cellular levels of plant EPSP by insertion of extra copies of the plant gene, or to introduce into sensitive cells a glyphosate-resistant EPSP and so allow plant growth to continue in the presence of normally toxic amounts of the herbicide. Although such strategies have been successful in creating herbicide-tolerant transgenic cultivars, insertion of the EPSP synthase genes into the plant cell nucleus unfortunately means that the novel genes can spread to other neighbouring crops, or wild relatives, through the movement of pollen. Not surprisingly, this has become a matter of great concern not only to the agricultural industry and environmentalists, but also to the public at large. Many problems could be initiated by gene flow to wild relatives and specifically through the creation of herbicide-tolerant weeds. To circumvent such criticism, a recent development has been to target the EPSP synthase gene directly into the chloroplast genome of tobacco plants. With this novel strategy, not only has it been possible to obtain very high levels of transgene expression, but also, because chloroplasts are inherited maternally in many species, it should prevent transmission of the gene by pollen. In short, one of the potential escape routes for certain transgenes into the environment has been sealed, although other exits still exist, for instance paternal or bipaternal inheritance of chloroplasts is common in gymnosperms and in some types of angiosperms.

10.1.7 GENETICALLY MODIFIED ORGANISMS AND SAFETY

In general it is now important that resources be directed towards research that will enable a better assessment of the ecological, toxicological and economic risks associated with *genetically modified organisms (GMOs)*, such as the consequences of the induced changes in plant metabolism and gene flows. Science has a key role to play in establishing public policies and implementing them, as well as creating conditions for their social acceptance.

10.2 Biomarkers of Environmental Stress

There is a growing interest in the measurement of stress proteins as novel biomarkers. Biomarkers are used in environmental monitoring to provide an early quantitative indication that organisms have been exposed to, or adversely affected by, environmental toxins. Ideally, effective biomarkers must have biological relevance and should be more sensitive than cruder toxicological end-points in present use such as growth, survival or ability to reproduce. In addition, they should be easy to detect and measure in a wide variety of ecologically relevant organisms, to provide a rapid indication of sub-lethal levels of toxicants or environmental stress.

10.2.1 CANDIDATE STRESS PROTEINS

Probably the most abundant and widely studied group of stress proteins that have potential as biomarkers is the *Hsp70 family* (see section 9.1), which are induced by a wide range of agents and conditions that either damage cellular proteins directly, or damage them indirectly by causing the production of abnormal proteins within cells. In addition to heat, stressors include heavy metals, anoxia and oxidizing agents. Hsp60s, because of their particular mitochondrial location, may serve as useful markers of specific effects at the level of mitochondria. Other stress proteins that have attracted attention as possible biomarkers are the *metallothioneins* and the *cytochrome P450s*. The latter group can transform many relatively insoluble organic compounds including drugs, pesticides, and polycyclic aromatic or halogenated hydrocarbons into water-soluble forms that can be excreted in bile or urine. Unfortunately, these enzymes can also convert some compounds (e.g. benzo[*a*]pyrene) to highly toxic forms capable of forming adducts with DNA which can bring about mutations and possibly cancer (see Chapters 2 and 8). Many of these cytochrome P450s can be induced within cells following exposure to their substrates or related compounds. The metallothioneins can chelate heavy metal ions (such as Cd^{2+}, Hg^{2+}, Zn^{2+} and Cu^{2+}) and are believed to protect mammalian cells against free heavy metal ion damage, although their precise role in cellular metabolism is yet to be fully understood. Their potential as biomarkers of toxic metal ion exposure lies in the fact that their intracellular levels increase significantly following exposures to a range of metal ions that depends on the species. A possible drawback is that they can also be induced by glucocorticoids, and physiological and nutritional conditions may affect their expression. *ATP-dependent efflux pumps*, such as the multidrug resistance transporter, are distributed throughout the animal kingdom, and may be also be useful as biomarkers.

10.2.2 APPLICATIONS

Biomarkers of stress have been employed in the comparison of whole animals in the wild from polluted and non-polluted sites. For example, the changes in liver levels of cytochrome P450 (CYP1A) expression can be monitored in different species of fish over extended periods, at various sites, with widely disparate levels of chemical contamination. Alternatively, caged fish can be maintained at contaminated sites and then compared with identical animals maintained under 'clean' laboratory conditions. In addition to P450s, the determination of comparative levels of heat-shock protein, or metallothionein, gene expression in fish and marine invertebrates of fresh water or salt water environments also has potential as an informative indicator of sublethal levels of toxicants or environmental stress.

As an alternative to the direct use of whole animals, primary cell cultures derived from relevant organisms offer a more rapid and economical approach for screening potential environmental stressors and for testing samples taken from suspected polluted sites. An example is a rat-derived *hepatoma cell line, H4IIE*, in which the induction of specific P450 activity is used as a bioassay for detecting planar halogenated hydrocarbons in environmentally derived samples. To avoid the problem of complex enzyme assays in the determination of biomarker levels, it has been possible to genetically engineer cell cultures to express a range of reporter genes. For example, animal cell culture systems carrying *bacterial lacZ* or *chloramphenicol acetyltransferase* 'reporter' genes driven by *stress-inducible promoters* can be developed and the level of stress-induced reporter gene product assayed immunologically. In a variation of this approach, it has been possible to place mammalian cell promoter sequences that respond to aromatic hydrocarbons and dioxins (i.e. the *XREs, or xenobiotic responsive elements*, see section 9.8) upstream of a *firefly luciferase gene* and transfer this genetic construct into the human *hepatoma cell line, HepG2*. This results in a highly sensitive bioluminescent assay for pollutants such as dioxin and aromatic hydrocarbons.

Although cell cultures are well suited for use in testing defined chemical compounds, they can be sensitive to toxic side-effects that may mask the activity of the reporter gene. Cell culture reporter-gene systems are also likely to be unsuitable for the direct testing of soils and sediments. A rapid and cost-effective *in vivo* bioassay system has been developed using transgenic strains of the soil nematode, *Caenorhabditis elegans*. These animals carry various fusions of the nematode *hsp16 genes* together with the *E. coli lacZ gene*. After exposure to stressor, the worms are permeabilized with acetone and are then assayed directly for β-*galactosidase* activity, using either colorimetric or histological procedures. Nematodes are easy and inexpensive to rear and can be directly exposed to soil or sediment samples as well as liquids.

The ability of free-living organisms to tolerate a wide range of growth conditions allows a considerable flexibility in dosage, especially when testing crude water samples of chemical mixtures. In addition to the nematodes, transgenic mice models are also being developed for the analysis of toxic inorganic compounds. In these animals, the *human growth hormone gene* has been placed under the control of the *human hsp70 gene promoter*, and intraperitoneal injection with sodium arsenite, cadmium chloride, copper sulphate or methyl mercurium chloride has elicited significant levels of human growth hormone in the plasma of these animals.

For such techniques to be effective it will be necessary to link levels of biomarker expression directly with adverse physiological events at both the cellular and organismal levels. With this combined strategy it will be possible to supplement traditional biological, chemical and physical methods for monitoring environment toxicity.

10.3 *Genes from Organisms Living in Extreme Environments*

Many microorganisms have been found to thrive in what appear to be harsh environments. Most of these organisms have been dubbed *extremophiles* and belong to the lineage *archaea*. While they lack nuclei, they resemble bacteria in many ways. In addition to possessing genes that are similar to those found in bacteria, they also have genes, otherwise found only in eukaryotes, as well as some genes unique to archaea. It is now believed that the distinctive biochemistry of molecules, such as the enzymes and other proteins, within these organisms that allows them to function in extremely hostile environments. If isolated and purified in sufficient quantities, such stress-resistant enzymes from extremophiles have commercial applications in environmentally clean industrial chemical processes that would capitalize not only on their precise specificity as biocatalysts, but also or their ability to tolerate high temperatures, or extremes of pH.

Key questions relate to the means whereby the proteins of extremophiles remain active under conditions that would be lethal for most organisms. Although many enzymes from *thermophiles* are quite similar to their counterparts in heat-sensitive organisms, they often have more ionic bonds and other internal stabilizing forces. The heat insensitivity of enzymes from *hyperthermophiles* has already led to a number of commercially significant industrial applications. One that is particularly relevant to the progress of molecular biology is the *Taq polymerase* which has been isolated from *Thermus aquaticus* and is used extensively in *in vitro* DNA amplification techniques

TABLE 10.1 *Extremophiles and possible biotechnological applications*

Class of extremophile	Characteristics	Applications
Thermophiles	Grow at 48–80°C, in hot springs	
Hyperthermophiles (heat-loving)	Grow at 80–150°C, in hot springs, and undersea hydrothermal vents	Heat-stable enzymes, e.g. *Taq* and *Pfu* polymerases for PCR technologies
Psychrophiles (cold-loving)	Grow at 4–12°C, e.g. Antarctic sea ice/ocean water	Enzymes that function at refrigerator temperatures, e.g. in food processing, cold wash detergents
Acidophiles (acid-loving)	Grow at pH 1 or less, in hydrothermal vents and some hot springs	Enzymes for syntheses in acid conditions, animal feeds
Alkaliphiles (alkali-loving)	Grow at around pH 8, in carbonate-laden soils and soda lakes	Enzymes for use in alkali-based laundry detergents
Halophiles (salt-loving)	Grow in high saline, e.g. salt lakes and solar evaporation ponds	Enzymes for syntheses in high salt environments

using the *polymerase chain reaction (PCR)*. In this technique the DNA polymerase is used to copy repeatedly a fragment of DNA in reactions that are cycled alternatively between high and low temperatures. It is the heat resistance of the Taq polymerase that has made it an ideal candidate for this cycling process. Another even more heat-resistant DNA polymerase is *Pfu polymerase*, which works best at 100°C and comes from *Pyrococcus furiosus*.

Table 10.1 gives characteristics of some of the best known types of extremophile, as well as some of the potential applications of enzymes isolated from these microbes. From the commercial point of view it is now more economical to produce these 'extremoenzymes' by genetic engineering rather than relying on large-scale culturing of the microbes themselves, which in light of their unusual characteristics could pose many technical difficulties.

Using recombinant DNA techniques, it is possible to isolate genes for extremoenzymes from small-scale extremophile cultures, and then insert these genes into ordinary bacteria, which will subsequently serve as a ready source of unlimited supplies of these enzymes in pure form.

10.4 *Bioremediation*

Bioremediation refers to the general processes whereby microbes (or other biological catalysts) act on pollutants to remedy or eliminate environmental contamination. This includes the reduction of organic chemical waste content of soils, ground water and effluents from food processing and chemical plants. In its simplest form, bioremediation relies on the indigenous microbial flora of soils or ground water. The assumption is that the microorganisms already at the site are already adapted to the particular chemical wastes, and are able to degrade these wastes effectively although supplies of oxygen, nitrates and phosphates may be limiting.

Microorganisms play a particularly fundamental role in the global recycling of matter, mainly due to their metabolic versatility. In particular, they excel at utilizing organic substances, natural or synthetic, as sources of energy and nutrition. Microorganisms have coexisted for billions of years with an immense variety of organic compounds. In addition to the organic components that make up living organisms, the natural world also contains a diversity of organic compounds produced by living organisms (which even includes organohalogen compounds), as well as compounds created by biogeochemical processes such as those in coal, petroleum and natural gas production. For example, petroleum oils originate from accumulation of phytoplankton residues in shallow seas and, although they are complex, their major constituents are paraffins (such as n-heptane, n-hexane, 2-methylpentane and 2,3-dimethylhexane), cycloparaffins (such as cyclohexane, methylcyclopentane and trimethylcyclopentane) and aromatics (such as toluene and benzene). Coal, on the other hand, is mainly derived from terrestrial plant matter, and is made up of aromatic rings fused into different polycyclic clusters that are linked to aliphatic clusters. The aromatic rings are variously substituted with phenolic, hydroxyl, quinone and methyl groups. Additionally, coal contains a variety of sulphur compounds. Over many many years, sources like these have provided microbes with a vast diversity of potential substrates for growth. This has resulted in the evolution of many enzyme types capable of

metabolizing widely disparate natural organic compounds by different catalytic routes. Such a repository of microbial enzymes now serves as a very effective platform for further evolutionary steps to permit the microbial utilization of many new synthetic compounds as they arise in the environment due to human activity.

Natural microbial communities are complex groups of highly interdependent microorganisms. Organic compounds are usually more effectively degraded in environments containing an array of microorganisms rather than in cultures containing single organisms, since the range of metabolic possibilities is likely to be greater in a community. Moreover, metabolic products produced by one microorganism in a community are quite likely to serve as substrates for others. Overall, the combined activity of several organisms may lead to the complete degradation of the contaminant to carbon dioxide, water and minerals.

An important application of microbial metabolic versatility is in the treatment of sewage and waste water which exploits the degradative capabilities of various bacteria and fungi to remove organic matter. In section 9.9, the ability of soil Pseudomonads to utilize a wide range of organic compound as carbon and energy sources was described. Many of these compounds are catabolized by enzymes encoded by plasmid-borne genes within these bacteria (see Table 9.5). In soil, there are many groups of these bacteria, each carrying a plasmid peculiar to the catabolism of a certain type of compound. As these plasmids are transmissible, the exploitation of a wide range of nutritional possibilities and responses to potentially toxic organic compounds through *gene sharing* is possible.

Despite the ability of microrganisms to utilize an enormous range of compounds produced in biological and geochemical processes, there are now many synthetic compounds, produced for industrial and agricultural purposes, that are released into the environment, but have no clear relationship with compounds already in the natural world. These compounds are generally referred to as *xenobiotics* and unfortunately can often be extremely stable in the environment. They include halogenated aliphatic and aromatic compounds, nitroaromatics, phthalate esters and polycyclic aromatic hydrocarbons. Many are distributed as components of fertilizers, herbicides and pesticides. Others such as polycyclic aromatic hydrocarbons, dibenzo-*p*-dioxins and dibenzofurans, are products of combustion reactions. Other, xenobiotics are detectable in waste effluents resulting from the production and use of synthetic products. A hazardous aspect of a number of these xenobiotics is that they can become progressively more concentrated with each step in the food chain. Striking examples are PCBs (polychlorobiphenyls) and phthalate esters (plasticizers for cellulose and vinyl plastics), which can act to disrupt the correct gene activation programmes of animal hormones (see section 9.8). PCBs are prepared by the partial chlorination of biphenyl and have a wide range of applications including use as hydraulic fluids, plasticizers, lubricants, flame retardants and dielectric fluids. Although their manufacture was halted in 1977, their degradation in the environment has been extremely slow.

Since there is already an enormous range of organisms capable of degrading a variety of environmental organic compounds, a pertinent question now is whether the techniques of genetic engineering can be used to improve the control of pollution. In principle it would be possible to manipulate genes to modify the specificities or catalytic capabilities of existing degradative enzymes; or to modify gene regulatory circuits to increase the range of degradative possibilities; or to create, in one microorganism, an assemblage of 'hand-picked' relevant degradative enzymes from widely disparate species. However, it is possible that such a laboratory-engineered 'super bug' may have a limited capacity for survival in the real environment, in the face of potentially novel adverse conditions, or from the competition of existing soil microflora.

Phytoremediation is a natural bioremediation process carried out by plants, especially those which have been able to survive in contaminated soils and water. Plants that are able to absorb high levels of contaminants with their roots and concentrate them there or in leaves and shoots are referred to as *hyperaccumulators*. Such plant species are found in areas where the soil contains greater than normal amounts of toxic metals as a result of pollution or geological factors. For instance, plants from the genus *Thlaspi* (Alpine pennycress) that can accumulate zinc, cadmium or lead, and *Assylum* species which accumulate nickel, have been used in soil bioremediation field trials. Plants from many other families have also been shown to remove cobalt, copper, chromium, manganese or selenium from contaminated soils. The ideal hyperaccumulators should be able to accumulate several different metals and function effectively even where the environmental concentration of the contaminants is relatively low, as well as being resistant to diseases and pests.

In addition to hyperaccumulation, *phytovolatilization* is being considered. In this process the objective is to turn the absorbed metal contaminants into gases. Rice, broccoli, and cabbage have an especially high selenium volatilization rate (i.e. conversion of selenium to dimethylselenide and dimethylselenide). Genetic engineering has recently been used to transfer a bacterial stress gene, encoding *mercuric reductase* (see Chapter 9.4) into *Arabidopsis* plants. When expressed in *Arabidopsis*, the product of this gene reduces mercury salts to volatile Hg^{0}.

10.5 *Literature Sources and Additional Reading*

ALI, A., KRONE, P.H., PEARSON, D.S. and HEIKKILA, J.J., 1996, Evaluation of stress-inducible *hsp90* gene expression as a potential molecular biomarker in *Xenopus laevis*, *Cell Stress and Chaperones*, 1, 62–69.

ALLEN, R.D., WEBB, R.P. and SCHAKE, S.A., 1997, Use of transgenic plants to study antioxidant defenses, *Free Radical Biology and Medicine*, 23, 473–79.

ATTFIELD, P.V., 1997, Stress tolerance: the key to effective strains of industrial baker's yeast, *Nature Biotechnology*, 15, 1351–57, New York: W.H. Freeman.

CANDIDO, E.P.M. and JONES, D., 1996, Transgenic *Caenorhabditis elegans* strains as biosensors, *Trends in Biotechnology*, 14, 125–29.

COLLIER, T.K., ANULACION, B., STEIN, J.E., GOKSOYR, A. and VARNASHI, U., 1995, A field evaluation of cytochrome P4501A as a biomarker of contaminant exposure in three species of flat fish, *Environmental Toxicological Chemistry*, 14, 143–52.

DANIELL, H., DATTA, R., VARMA, S., GRAY, S. and LEE, S.-B., 1998, Containment of herbicide resistance through genetic engineering of the chloroplast genome, *Nature Biotechnology*, 16, 345–48.

DE LA FUENTE, J.M., RAMIREZ-RODRIGUEZ, V., CABRERA-PONCE, J.L. and HERRERA-ESTRELLA, L., 1997, Aluminium tolerance in transgenic plants by alteration of citrate synthesis, *Science*, 276, 1566–68.

FISCHBACH, M., SABBIONI, E. and BROMLEY, P., 1993, Induction of the human growth hormone gene placed under human hsp70 promoter control in mouse cells: a quantitative indicator of metal toxicity, *Cell Biological Toxicology*, 9, 177–88.

GLAZER, A.N. and NIKAIDO, H., 1995, *Microbial Biotechnology, Fundamentals of Applied Microbiology*, pp. 561–620, Basingstoke: W.H. Freeman.

HAASCH, M.L., PRINCE, R., WEJKSNORA, P.J., COOPER, K.R. and LECH, J.J., 1993, Caged and wild fish: induction of hepatic cytochrome P450(CYP1A1) as an environmental biomarker, *Environmental Toxicological Chemistry*, 12, 885–95.

HUGHES, M.A. and DUNN, M.A., 1996, The molecular biology of plant acclimation to low temperature, *Journal of Experimental Botany*, 47, 291–305.

INGRAMS, J. and BARTELS, D., 1996, The molecular basis of dehydration tolerance in plants, *Annual Review of Plant Physiology and Plant Molecular Biology*, 47, 377–403.

JAGLO-OTTOSEN, K.R., GILMOUR, S.J., ZARKA, D.G., SCHABENBERGER, O. and THOMASHOW, M.F., 1998, *Arabidopsis CBF1* overexpression induces *COR* genes and enhances freezing tolerance, *Science*, 280, 104–06.

JORGENSEN, R.B., HAUSER, T., MIKKELSEN, T.R. and OSTERGARD, H., 1996, Transfer of engineered genes from crop to wild plants, *Trends in Plant Science*, 1, 356–58.

McCUE, K.F. and HANSON, A.D., 1990, Drought and salt tolerance: towards understanding and application, *Trends in Biotechnology*, 8, 358–62.

MADIGAN, M.T. and MARRS, B.L., 1997, Extremophiles, *Scientific American*, April, 66–71.

MOUSDALE, D.M. and COGGINS, J.R., 1991, Amino acid synthesis, in Kirkwood, R.C. (Ed.), *Target Sites for Herbicide Action*, pp. 29–56, New York: Plenum Press.

MURATA, N., ISHIZAKI-NISHIZAWA, O., HIGASHI, S., HAYASHI, H., TASAKA, Y. and NISHIDA, I., 1992, Genetically engineered alteration in the chilling sensitivity of plants, *Nature*, 356, 710–13.

ROGERS, H.J. and PARKES, H.C., 1995, Transgenic plants and the environment, *Journal of Experimental Botany*, 46, 467–88.

RUGH, C.L., SENECOFF, J.F., MEAGHER, R.R. and MERKLE, S.A., 1998, Development of transgenic yellow poplar for phytoremediation, *Nature Biotechnology*, 16, 925–8.

RYAN, J.A. and HIGHTOWER, L.E., 1996, Stress proteins as molecular biomarkers for environmental toxicology, in Fiege, U., Morimoto, R.I., Yahara, I. and Polla, B.S. (Eds) *Stress-inducible Cellular Responses*, pp. 411–24, Basel: Birkhauser Verlag.

SACCO, M.G., ZECCA, L., CHIESA, G., PAROLINI, C., BROMLEY, P., CATO, E.M., RONCUCCI, R., CLERICI, L.A. and VEZZONI, P., 1997, A transgenic mouse model for the detection of cellular stress induced by toxic inorganic compounds, *Nature Biotechnology*, 15, 1392–97.

SALT, D.E., SMITH, R.D. and RASKIN, I., 1998, Phytoremediation, *Annual Review of Plant Physiology and Plant Molecular Biology*, 49, 643–68.

SONG, W.-Y., WANG, G.-L., CHEN, L.-L., KIM, H.-S., PI, L.-Y., HOLSTE, T., GARDNER, J., WANG, B., ZHAI, W.-X., ZHU, L.-H., FAQUET, C. and RONALD, P., 1995, A receptor kinase-like protein encoded by the rice disease resistance gene, *Xa21*, *Science*, 270, 1804–806.

TILLIT, D.A., ANKLEY, G.T., VERBRUGGE, D.A., GIESY, J.P., LUDWIG, J.P. and KUBIAK, J.T., 1991, H4IIE rat hepatoma cell bioassay-derived 2,3,7,8-tetrachlorodibenzo-*p*-dioxin equivelants in colonial fish-eating waterbird eggs from the Great Lakes, *Archives of Environmental Contamination Toxocology*, 21, 91–101

WATANABE, M.Y., 1997, Phytoremediation on the brink of commercialization, *Environmental Science and Technology*, 31, 182A–186A.

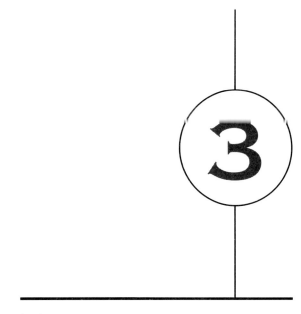

THE ENVIRONMENT, GENES, BIODIVERSITY AND CANCER

CHAPTER 11

Prospects

■ Increasingly precise knowledge concerning the nature and consequences of gene damage, mutation and gene activation
caused by specific environmental factors is likely to form the
basis of more effective future strategies to reduce environmental
hazards as well as the incidence of cancer.

■ Some environmental factors may also influence the manner
in which certain genes are inherited through epigenetic
mechanisms.

■ Certain environmental stress proteins may have profound
effects on evolutionary outcomes as a result of their ability
to influence development and morphogenesis.

■ A 'gene bank' of biodiverse stress genes could be a valuable
asset for future genetic engineering of organisms to tolerate
increasing environmental adversity.

In recent times there have been a variety of attempts to estimate the number of species on earth. Although around 1.5 to 1.8 million species have already been described, it may be that the total number of species lies somewhere in the region of 5 million. This could be a conservative estimate, and as many as 30 million has been suggested.

Along with these attempts to gauge the present extent of biodiversity, efforts are being made to evaluate likely rates of species extinction. On the basis of present global deforestation rates of around ten million hectares per year, it is believed that tropical forest species will become extinct on a global scale at a rate of 4–8% over the next 25 years. Since ecologists estimate that 50–90% of the earth's species inhabit tropical forests, this means that around 4000 to 14 000 species may be lost each year. In contrast, examination of the fossil record, although indicating that the majority of species that have existed on Earth are now extinct, suggests an average background extinction rate of one to five species per year. On the face of it, species are being lost at a rate that is many times more rapid than at any previous time. Although man is the primary driving force behind these extinctions, through efforts to modify the natural environment to create new land for agriculture, industry and dwellings, it is only recently that the value and magnitude of biodiversity has been recognized. Man is now extremely dependent on the biodiversity of species for a variety of essential services, including the recycling of atmospheric gases, water and nutrients.

This chapter aims to provide an overview of genes in the contexts of biodiversity and environmental adversity, as well as the prospects of environmentally-induced DNA damage and gene mutation in relation to human diseases such as cancer.

11.1 *Genes and Biodiversity*

It can be argued that the fundamental currency of biodiversity is individual genes, as species are essentially characterized by differences in their genes. In terms of their potential to adapt to new environments, species require a diversity of genes within their population, so that genes that might be rare and of limited use in certain environments may be valuable for survival in some novel environment. The potential of an organism is determined by its genes, but genes do not exist in an environmental vacuum; neither do they function without reference to their fellow genes. The activity of genes can be influenced by the internal environment of the organism as well as by the factors in the external environment A single genotype may give rise to many different phenotypes under varying circumstances of environment. In turn, one environmental factor may act to elicit various responses depending on the genotype of the organism which determines the threshold of the responses.

Humans have the ability to alter the landscape drastically, a feature which distinguishes them from most other species. For example, it has been estimated that roughly a quarter of the carbon dioxide released into the atmosphere at the moment arises as a result of the transformation of tropical forests into pasture. Such ecosytems are important for the maintenance of the composition of the atmosphere. As levels of atmospheric carbon dioxide increase a greenhouse effect is produced, which is likely to have long-term effects on global climate that could profoundly alter the weather conditions across different regions. Plant and animal populations, if they are unable to migrate rapidly enough to survive, will have to adapt to new conditions as they arise in their particular geographical location. The processes of natural selection will propel these populations in the direction of appropriate adaptation, but if the climatic change is too abrupt, adaptation may be beyond the limits of these selective mechanisms. Overall, the effectiveness of a population will depend on its level of genetic variability. In the event, the species best able to deal with climatic changes are those that can both migrate and adapt. They will have large populations and wide geographical ranges. In contrast, climatic changes induced by global warming may well present considerable threats to the survival of small populations of endangered species.

11.2 *Genes for Environmental Adversity*

Most organisms can adjust their spectrum of active genes in the face of environmental adversity. A range of environmental threats, or stresses, can lead to the rapid production of specific cellular stress proteins that appear to play crucial functions in helping the organism to resist particular threats (see Chapter 9). There is intense research activity aimed at elucidating the molecular mechanisms whereby specific environmental adversity is sensed and how the activation of relevant stress genes is triggered.

While the ability of organisms to adjust the spectrum of active genes has significance for the processes of acclimation to short-term changes in environmental conditions, the possession of a diversity of genes that encode potentially protective stress proteins may have crucial adaptive evolutionary significance in the long term. Variants of such genes could be selectively advantageous when populations of organisms are confronted with more long-term environmental changes that may be variously stressful or hostile. The appreciation and

understanding of the diversity of genes in all types of organism encoding such protective proteins and the mechanisms that regulate their expression are therefore important aspects of biodiversity. A biodiverse range of stress genes, governing graded responses to a range of environmental threats, is a future asset. It would provide the basis of a valuable gene bank, possibly containing many unique variants of stress genes, that could provide opportunities to 'engineer' genetically particular chosen organisms to tolerate increased environmental adversity. An example has been the transfer of a pathogen-resistance gene (encoding a specific plasma membrane pathogen elicitor receptor) from a resistant wild strain of rice to a domesticated pathogen-sensitive strain (see Chapter 10). Initially, such approaches may be particularly important for the future survival of economically important species, for instance in agriculture and food production, but they may have wider application. Unfortunately, once a species is lost from the wild, the opportunity to understand its particular stress genes is gone. It is paradoxical that as biodiversity is under threat, our ability to appreciate the detailed nucleotide sequence organization of genes and whole genomes is increasing (driven by the Human Genome Project and the sequencing of genomes from certain bacteria, *Drosophila*, *Caenorhabditis* and *Arabidopsis*).

11.3 *The Environment, Gene Damage and Mutation*

Although some stressful environmental conditions can activate protective stress genes, other environmental factors are potentially toxic. At sub-lethal levels they can have effects which can be far-reaching and long-term, particularly their capacity to bring about heritable alterations in the chemical structure of genes. A considerable amount of effort has gone into analyzing the precise nature of the chemical damage that such environmental factors can cause, either directly or indirectly, to the base, sugar and phosphate components of DNA molecules (see Chapter 2). Examples of well studied factors include many of the synthetic chemicals associated with industrialized communities. These include the hazardous air pollutants that arise from tobacco smoke, automobile exhaust and various industrial sources; ultraviolet radiation, which is likely to increase in significance as a result of the shrinking ozone layer; the ionizing radiation which will always be a risk in relation to energy generation; and the natural and synthetic contaminants of food. The damage to DNA can be quite varied, as described in Chapter 2. Such 'environmental' damage is superimposed on the considerable level of 'background' DNA alteration that takes place spontaneously and continuously within living cells (see Chapter 3). Cellular DNA is also vulnerable to the oxidatively damaging effects of reactive oxygen species continuously generated intracellularly as by-products of normal energy metabolism, or in mammals from cells of the immune system (see Chapter 4).

Despite the opportunity for extensive DNA damage arising from environmental factors, organisms are equipped with a variety of enzyme systems which can reduce such molecular hazards through various metabolic strategies (see Chapter 5). The existence of multiple alleles at the genetic loci that encode these detoxification enzymes can result in differential susceptibilities of individuals within a population to the DNA-damaging effects of many environmental chemicals and drugs.

Organisms are also furnished with an array of antioxidant systems to limit the potential for free radical attack to cellular structures including DNA (see Chapter 5). These cellular antioxidant systems contribute to the reduction of hazardous free radicals either generated within cells or from external environmental sources such as ionizing radiation, strong sunlight, or pollutants such as cigarette smoke and automobile exhaust fumes. Much recent health and nutritional interest stems from the fact that man can reinforce reserves of many of the small molecular weight antioxidants, from dietary fruits and vegetables. Absence of fruit and vegetables from the diet is thought to make DNA, and other cellular structures, more vulnerable to free radical attack.

This impressive array of cellular protection systems may still be overwhelmed in certain severe environmental conditions, thus placing the structure of genes at risk. As a further safeguard, another layer of protection exists in the shape of enzyme systems that can repair DNA damage in various ways (see Chapter 6). In any one organism there are a number of distinct DNA repair systems available to repair different types of damage. Each system, however, is made up of a considerable number of individual gene products, and it is perhaps not surprising that defects in genes encoding such products can prejudice the efficiency of repair processes. Indeed, a number of human diseases are characterized by defects in the capacity to repair DNA damage.

In proliferating cells an adequate response to DNA damage nevertheless requires more than just the repair of the DNA lesion (see Chapter 8). In animal cells, 'checkpoint' controls arrest the cell division cycle after damage to DNA, allowing repair to DNA to take place before alterations are perpetuated as a result of ensuing DNA replication processes. In multicellular organisms severe DNA damage in particular cells can induce them to follow biochemical pathways leading to programmed cell death or apoptosis, thus protecting the organism from individual cells that may be severely impaired.

DNA alterations that escape repair are perpetuated by replication as mutations (see Chapter 8). This will add to the extent of genetic variation in populations of organisms which, in turn, provides the raw material for natural selection. Sophisticated molecular biological methods are now available to detect and locate mutations in specific genes. Analysis of the spectrum of location of mutational 'hot spots' will provide a means of establishing precise relationships between different environmental factors and heritable genetic change. Most mutations caused by environmental factors thus have well-understood physicochemical origins, and, although often thought to be random, are more likely to occur in some genes than in others. Evolution has fine-tuned most organisms and thus it might be expected that most mutations are disadvantageous. Nevertheless, it could be argued that environmentally-induced mechanisms that produce a larger selection of mutants would provide a greater chance of producing at least one descendant better equipped to cope with the new crisis.

As already mentioned, there are genes whose encoded products are intimately involved not only in the protection of DNA but also in the fidelity of copying and repair of genes. These are themselves vulnerable to mutation from environmental factors, and so overall mutation rates could also be influenced by this indirect route. Such mutated genes act as 'mutators' by increasing the mutation rate and again may help an organism to acquire that rare mutation which is an improvement. The mutator gene will then be part of the genetic make-up of the organism it helped to create. However, in evolutionary terms this association is likely to be short-lived, as it may be lost during the sexual mixing of genes that occurs at ensuing reproductive cycles. Overall, any

environmentally-induced mutator-type genes will tend to disappear from the population, and thus any increases in mutation rate will also decline. In the long term, the effects of a mutator gene are likely to be disadvantageous despite the occasional beneficial mutation. In contrast, in organisms reproducing asexually, mutator genes will not become dissociated from any advantageous genes they produce, and thus could initiate the propagation of increasingly successful organisms.

In the context of possible genetic improvement, microorganisms also regularly swap genes by horizontal DNA transfer mechanisms which almost seem to cheat evolution. They can take up DNA by various means that may provide them with new abilities faster than would be possible by mutation. The mechanisms include bacteriophage transduction, plasmid transfer by conjugation and direct DNA uptake by transformation. Plasmids are important, because they harbour a range of genes that encode, for example, virulence factors, toxins, catabolic enzymes and antibiotic resistance factors. Plasmid transfer between bacterial species is quite common, but transfer can also occur between kingdoms, such as from bacteria to plants and yeasts. DNA uptake by transformation, on the other hand, appears to be limited to only a few bacterial species and occurs under somewhat limited circumstances. *Bacillus* and *Streptococcus* can take up any type of DNA provided they first are rendered competent. *Haemophilus* can only take up DNA fragments provided they have a certain 11 base-pair recognition sequence that normally occurs frequently in the *Haemophilus* genome. Overall, the majority of successful transformation events are intraspecific and, while transformation could simply be a means of repairing mutational damage, a key question is whether transformation occurs significantly in the environment. Given the fact that evolution has proceeded in part by mutation and in part by recombination, it is possible that transformation could have evolved in organisms that lack other mechanisms of exchanging DNA. What is necessary in the environment is the release of DNA from cells and the generation of competent bacteria. Many Gram-positive bacteria such as *Bacillus* do release DNA as they age, or if they are subjected to starvation or deleterious environmental conditions. Thus there would be a ready supply of DNA in the environment and, although an infrequent event, transformation may be an important force in microbial evolution. It has been suggested that entire microbial populations may share common gene pools of extrachromosomal elements. When combined with intracellular mechanisms for DNA rearrangement, including those involving transposons, the whole genomic repertoire of a particular bacterium is theoretically available to other members of the bacterial community through conjugation, transduction and transformation.

11.4 *Genes, Cancer and the Environment*

Mutational events caused by certain environmental agents are also of immediate concern in terms of the health of humans and other animals. For example, there are genes in somatic cells which are important because their protein products are involved in the proper regulation of the normal processes of cell division. If these particular genes become mutated, this can sometimes have severe consequences, giving rise to altered proteins whose ability to control the growth of cells correctly is compromised. One outcome of deregulated cell division is the development of cancers. Mutated somatic genes may also

contribute to other degenerative human diseases and to the ageing process.

From an environmental viewpoint, cancer is a serious consideration. The disease is now striking the general population with increasing frequency. In the USA some 40% of the population will develop cancers. At least 80% of those are likely to be due to environmental factors whose effects could be either eradicated or at least reduced. Industrialized countries now appear to have disproportionately more cancers than countries with few industries, and migrant studies indicate that it is the new environment that determines the cancer risk rather than genetic background. Since 1950, although there has been little change in survival rate, the overall incidence of cancers has risen by about 1% per annum. Particularly notable are the increases in lung cancer, multiple myeloma, melanoma, non-Hodgkin's lymphoma and breast cancer, although cancers of the prostate, testes, kidney and larynx have also risen. Although cancer is mostly thought of as a disease of old age, childhood leukaemias and other cancers have also shown disturbing increases (roughly 20% between 1950 and 1988).

The years since 1945 have seen a rapid increase in the rate at which novel synthetic compounds have been created. It has been estimated that there are now around 75 000 synthetic chemicals in common use, many of which are carcinogenic and against which man has no natural protection. Safety procedures and protective legislation have slowly evolved for the workplace. Unfortunately, it is not just industrial workers that are exposed. Such massive amounts of chemicals are now produced and used that that they are no longer contained in the workplace, but slowly seep into the environment due to their release into air and water, for instance from toxic wastes, accidental chemical spills, or farm run-off. Many environmental carcinogens, such as PCBs and pesticide residues, find their way into food and, regrettably, low levels of such synthetic carcinogens are now basic contaminants of our food. Moreover, children tend to receive higher doses of any chemicals present in air, water or food because they breathe, eat and drink proportionately more than adults.

Some estimates suggest that around 40 possible carcinogens contaminate drinking water, while 60 or so are released from industrial sources into the air and approximately the same number are routinely sprayed on food crops as pesticides. Organochlorides, many of which function as hormone disrupters, are highly persistent in air and water, for example DDT has a half-life of 7 years. They become deposited as a result of winds and storms in the soil and on vegetation and from there enter the food chain, with diet probably being the main route of exposure. Although DDT and PCBs were banned in the 1970s, both can still be detected in human tissues. By 1951, DDT-contaminated human breast milk was passing from mother to child. In humans, DDT is metabolized to DDE and an accumulation with age has been observed, particularly in fatty tissues. Breast cancer patients have often been found to have higher levels of DDE and PCBs in their tumours than in surrounding normal tissues, and both DDT and PCBs are linked with breast cancers in rodents.

The atmosphere is probably the largest receptacle for general industrial emissions. Although ostensibly reducing waste, incinerators are another source of contamination releasing dioxins and furans. One of these, TCCD (2,3,7,8-tetrachlorodibenzo-*p*-dioxin), is particularly carcinogenic. Additional sources of air-borne carcinogens are automobile exhaust, tobacco smoke and the burning of fossil fuels such as coal and coke. They contribute aromatic hydrocarbons that include benzo[*a*]pyrene which can be inhaled and stick

to the skin. Within cells, benzo[*a*]pyrene is activated by cytochrome P450s to a form that can form bulky adducts with DNA bases (see Chapter 2). In turn, these can bring about mistakes in DNA copying at ensuing rounds of DNA replication. High levels of such DNA adducts can be found in people with lung cancer, compared with levels in those without the disease. However, depending on differences in the overall effectiveness of cytochrome P450s between individuals, susceptibility to benzo[*a*]pyrene will vary. Air can also carry radioactive material driven into the stratosphere following atmospheric testing of nuclear devices. In areas of high rainfall this can sometimes be brought down to earth and enter the food chain. This route may partly explain the build up of strontium-90 in milk and its accumulation in human tissues following periods of such testing.

Many animal studies have shown chemical–cancer links. In the wild, animals inhabiting contaminated environments have developed cancers, and in the laboratory many synthetic (and natural) compounds will induce cancers in experimental animals, particularly rodents. With the exception of tobacco smoke and lung cancer, precise links in humans have, however, been difficult to establish, although there is much to link skin cell cancers to excessive ultraviolet radiation exposure. There is speculation that phenoxyherbicides might be linked with non-Hodgkin's lymphoma, which has tripled in incidence since 1950. Multiple myeloma has also increased over this period and may be associated with exposure to a variety of chemicals including metals, rubber, paints, solvents and petroleum products, as well as ionizing radiation. Farmers and agricultural workers exposed to pesticides and herbicides also have higher rates of this condition. In the case of bladder cancers, there is reason to suspect cigarette smoking and exposure to aromatic amines such as aniline, benzidine, naphthylamine and o-toluidine. In the past, workers directly involved in dyestuffs or tyre production were most susceptible, but now these carcinogens are widely present in rivers, ground water and dump sites and the incidence of bladder cancer has increased some 10% in the USA between 1973 and 1991, although half of these cases may be due to cigarette smoking. In the cases of the aromatic amines, human cells are provided with some protection in the shape of N-acetyl transferases (see Chapter 5) which can catalyze their detoxification. Susceptibility to these carcinogens is therefore likely to be influenced by the extent of polymorphism in our N-acetyl transferase genes. Poor 'acetylaters' in the population may therefore be at most risk from bladder cancers brought about by aromatic amines.

As pointed out in Chapter 8, it appears that cancers can arise from single cells and are 'initiated' by mutations in certain specific genes. Although humans have around 100 000 genes, for the development of cancers the critical mutations are those occurring in protooncogenes or tumour suppressor genes that are central to the control of cell division. Although some of these DNA mutations, resulting in hyperactive oncogenes and inactive suppressors, can be inherited, the majority are acquired during the lifetime of an individual. A second component of the carcinogenic process is referred to as 'promotion'. This involves stimulatory mechanisms that selectively increase the number of cells carrying the critical mutations in protooncogenes or tumour suppressor genes. Promotion does not involve mutations, but encourages cells to divide by altering the spectrum of genes expressed. In certain situations the natural hormone oestrogen acts as a promoter, as do a number of synthetic organochlorides in experimental animals. The progress of tumour formation is often a lengthy process involving somewhere between three and seven

successive rounds of cancer gene mutation, followed by the selective promotion of mutated cell growth. The lag time between the first mutational event and the development of a full-blown tumour can be as long as 15 to 20 years. However, it may be that the duration of the lag reflects the number of mutated protooncogenes or supressor genes already inherited by an individual. There may also be differences that are characteristic of different tumour types. In the case of leukaemias developed after the Japanese atomic bomb blasts, the disease was apparent after five years. Given the central role of mutations it is perhaps not surprising that mutations in genes that result in increases in the overall mutation rate, also accelerate the development of cancers. For example, individuals with inherited disorders of DNA repair such as Xeroderma pigmentosum, Bloom's syndrome, Fanconi's anaemia, Ataxia-telangiectasia and hereditary non-polyposis cancer (see Chapter 8) have a higher than normal predisposition to cancer.

In summary, some carcinogens cause mutations and act as 'initiators' of carcinogenesis, and others that are not mutagenic act as 'promoters', while still others can act both as initiators and promoters. Additionally, some carcinogens that do not cause mutations can bring about chromosomal aberrations directly, by disabling the spindle fibre apparatus leading to the aneuploidy that is also a feature of tumour cells. The process of carcinogenesis is complex and our ability to evaluate chemicals and other environmental factors in this context requires a fuller understanding of the different cellular mechanisms involved. Because the complete process is usually quite lengthy, adequate testing of suspected human carcinogens is fraught with difficulties. Many substances are branded as carcinogens on the basis of tests carried out in experimental animals (usually rodents) over very short time periods, often with inappropriately high concentrations. While providing useful indicators of potential carcinogenicity, such animal tests fail to provide the necessary hard information regarding the dangers to humans faced with long-term, low-level exposures to hazardous substances, which also may accumulate differentially in various human tissues. Moreover, substances that are carcinogenic in rodents may not be carcinogenic for humans and vice versa, and potentially adverse effects may be ameliorated by polymorphisms in human genes encoding detoxifying enzymes and antioxidant defences. On top of this is the problem of simulating environmental reality. In most situations, exposure is not to a single carcinogen at a time, which is all that most present toxicity tests can adequately deal with. Over long periods, there may be exposure to cocktails of environmental carcinogens, some with the propensity to accumulate in specific tissues but not in others; some with the ability to act in concert with others to potentiate their harmful effects; and some that can be metabolized within the body to an unknown variety of potentially more carcinogenic substances. Some environmental carcinogens will act directly to mutate DNA, or function as promoters, while others may act indirectly, for example by bringing about an increased generation of free radicals, or other reactive oxygen species within cells, which in turn could inflict oxidative damage on cellular DNA. Such reactive oxygen species could arise as a result of the interference with electron transfer in mitochondria by the carcinogens, or as by-products of the metabolism of carcinogens by cytochrome P450s. A recent report shows that, in rats exposed to a cocktail of 15 different herbicides, there were very significant increases in oxidative DNA damage. Reactive oxygen species are also elevated in animal cells exposed to the fungal carcinogen aflatoxin

B_1, or to low-level radiation in the form of α-particles.

While the direct testing of possible carcinogens in animals and in various cell culture systems has limitations, epidemiology offers another approach. Epidemiologists attempt to identify factors that are common to cancer victims' histories and ways of life and evaluate them in the context of current biological understanding. Their aim is to decide whether exposure to certain factors or characteristics can cause or significantly increase the odds of particular cancers developing. This has thrown light on many of the environmental causes of cancer, and also has given estimates of the annual deaths attributable to each.

Many observations suggest that the increases in cancer incidence over the past 50 years may be due to the widespread introduction of many new synthetic chemicals into the workplace. This is more plausible than suggestions that the increases may be due to increased levels of mutant protooncogenes and tumour suppressor genes in the human population. Mutant genes would not spread so rapidly in human populations. While consumption of natural carcinogens in the diet is likely to have remained unchanged over that period, the presence of small amounts of synthetic carcinogens in food, air and water has increased.

On the face of it, there is there is a pressing need for a more rigorous evaluation and awareness of the increasing hazards of synthetic chemicals in the environment. Much has already been done to protect those in the workplace with appropriate legislation, but not enough for the public at large. Clearly these are complex issues, but too often the acquisition of understanding has been slow. By the time the hazards of a particular environmental chemical are understood, the process of tumour development in those exposed is usually well advanced. It is not surprising, therefore, that the possibility of having drugs available that might provide instant cures for cancers in advanced states is a very attractive dream for the victims.

For many synthetic carcinogens, eradication from the environment can only be a very long-term target. It is not a simple matter to avoid many of them, as it is to evade ultra-violet exposure or tobacco smoke. Preventive anti-cancer strategies such as fruits and veg-etables in the diet have been suggested, but these carry the risks of synthetic pesticide contamination. Although fruits and vegetables contain high levels of natural pesticides, and the level of synthetic pesticides is proportion-ately extremely small, it should be stressed that synthetic pesticides do not actually belong in plant-derived foods. Indeed, the presence of synthetic pesticides in fruit and vegetables is really a post-World War II development. Overall, while fruit and vegetable consump-tion can be protective against a number of can-cers, it is notable that this is less impressive against hormone-dependent cancers such as breast and prostate cancers. Other protective strategies against cancer have been developed which involve the development of drugs for 'chemointervention'. Here the aim is to induce within cells increased levels of detoxifying enzymes, such as glutathione S-transferase, in an attempt to eliminate envi-ronmental carcinogens before they can do their damage. However, considering the long-term nature of the carcinogenic process, these pro-tective drugs would also have to be taken on a long-term basis with the attendant risk of undesirable side-effects.

There is a general threat to life from the chemical pollution of the planet, but what are the least toxic solutions? These will include not just more adequate containment of car-cinogenic substances in the workplace, but also imaginative biodegradation programmes, possibly utilizing the full genetic software of microorganisms, together with the reengineering

of genes encoding degradative enzymes to extend their substrate specificity. Although it could be argued that cancer is likely to have little effect on evolutionary outcomes, since it still mainly affects those past reproductive age, the chemicals and other environmental hazards that cause germ-line mutations may well be significant, although in no readily pre-dictable way. In contrast, those environmental chemicals that cause widespread reproductive and developmental problems undoubtedly have ominous implications.

11.5 The Environment and Epigenetic Inheritance Mechanisms

Despite the accumulating knowledge of human genes and their respective 'control' sequences (see Chapter 7), it is now becoming apparent that their inheritance may be more complex than originally envisaged. Modulation of gene activity can be achieved by mechanisms superimposed on those influenced by primary DNA sequences. Moreover, it appears that her-itable states of transcriptional repression, or activation, can be influenced not only by the chromosomal context of the promoter (e.g. its accessibility), but also by covalent modifica-tions to DNA and chromosomal histones.

Recent evidence indicates that the enzymic acetylation by *histone acetylases* of specific lysine residues in histones can act as markers through which states of gene activity, or inactivity (see Chapter 7.5), can be main-tained from one cell generation to the next, in a number of higher organisms. In other situ-ations, certain genes are known to be active only if inherited from the father, and in other cases only if they come from the mother. It is not yet clear how the parental origin of these genes is 'marked', but clearly some sort of *'imprinting'* must persist through generations to inform the offspring cells which genes to 're-imprint'. Certain studies suggest that genes can be marked for shut-down by the pattern of DNA cytosine residues covalently modified with methyl groups, through the action of *DNA methylases*, yielding 5-methylcytosines (see Figure 3.9). Crucially, the addition of such methyl groups to cytosines in certain genes and their control sequences can block gene transcription. Methylation of cytosines can also facilitate the folding of DNA into tight inaccessible coils, through the action of DNA-binding proteins specific for methylated DNA sequences. While enzymic mechanisms exist for the inheritance of specific patterns of DNA methylation from one cell generation to the next, it is also possible to alter the patterns of DNA methylation. For instance, altered patterns of DNA methylation have been observed when the 'environment' of mammalian nuclei is changed. An example occurs in the nuclei of mouse embryo cells following their transplantation into enucleated eggs from different mouse strains. The possibility of environmental factors (including drugs) influencing gene activity through *heritable epigenetic mechanisms* is now a serious prospect, particularly in relation to the development and evolution of higher organisms.

11.6 *Environmental Stress Genes and Evolution*

Recent research suggests that the stress protein, Hsp90, may be an important link between environmental contingency and morphological evolution. As described in Chapter 9, a group of proteins induced by environmental stress, namely the *heat-shock proteins*, play an important physiological role in maintaining the correct folding of cellular proteins. If mutations are sustained in genes for one of these stress proteins, Hsp90, this can lead to substantial *morphological defects* in affected fruit flies.

At normal temperatures Hsp90 binds to cell proteins which are critically involved in the regulation of cell proliferation and embryonic development. Hsp90 is believed to stabilize these proteins in their native form so that they can respond properly to stimulation by components of the relevant cellular signal transduction pathways. After heat shock, however, Hsp90 can be diverted to the protection of other cell proteins. In fruit flies, mutations in both copies of the *hsp90* gene are lethal, so mutations can only be maintained in heterozygous flies. However, a proportion of these heterozygotes are found to be deformed, with a range of malformed structures including legs, wings, antennae, eyes and bristle patterns. Importantly, these deformed flies none the less survive to adulthood and are fertile. It is suggested that the delicate balance between signal transduction pathways that underpins development and morphogenesis is disturbed. In the presence of mutant versions of Hsp90, the developmental processes then become susceptible to abnormal, or non-native, variants of cell signalling proteins that would normally be stabilized by wild-type Hsp90 molecules. The normal integration of development is disturbed and rapid morphological changes result. This would be compatible with the fossil record, which contains many examples of apparently rapid changes in body structures, with intervals of relative stability.

11.7 *Postscript*

Living organisms do not simply confront an autonomous physical world. Rather, they make a significant contribution to the assembly of the environment. For example, every organism is constantly changing the world in which it lives simply by taking up materials and excreting others. When food is consumed, gases and solid waste are produced, which are, in turn, the materials for consumption by some other organisms. Likewise, bacteria also use up food material and excrete products which can be toxic. Thus, in common with all other living organisms, humans cannot live without changing the environment.

Unfortunately, we now have the power to change the world more rapidly and more comprehensively than other organisms. Our lifestyles are such that we could easily recreate the environment in a way that is potentially hostile to our own survival. Crucially, it could be argued that it is our own genes – like those of any other organism – that, by influencing our morphology, physiology and behavioural patterns, are contributing to the restructuring of the environment.

11.8 *Literature Sources and Additional Reading*

AMES, B.N. and GOLD, L.S., 1997, Environmental pollution, pesticides, and the prevention of cancer: misconceptions, *FASEB Journal*, 11, 1041–52.

CADBURY, D., 1997, *The Feminization of Nature*, London: Penguin Books.

CARSON, R., 1965, *Silent Spring*, London: Penguin Books.

CUNNINGHAM, W.P. and SAIGO, B.W., 1990, *Environmental Science, a Global Concern*, Dubuque, IA: W.C. Brown Publishers.

DOBSON, A.P., 1996, *Conservation and Biodiversity*, New York: Scientific American Library.

EPSTEIN, S., 1998, Winning the war against cancer? . . . Are they even fighting it? *The Ecologist*, 28, 69–80.

FIEGE, U., MORIMOTO, R.I., YAHARA, I. and POLLA, B.S., 1996, *Stress-inducible cellular responses*, Basel: Birkhauser Verlag.

GARRETT, L., 1994, *The Coming Plague, Newly Emerging Diseases in a World out of Balance*, London: Penguin Books.

GASSER, S.M., PARO, R., STEWART, F. and AASLAND, R., 1998, The genetics of epigenetics, *Cellular and Molecular Life Sciences*, 54, 1–5.

GOLDSMITH, Z., 1998, Cancer: a disease of industrialization, *The Ecologist*, 28, 93–99.

KING, R.J.B., 1996, *Cancer Biology*, Harlow: Addison Wesley Longman.

LODOVIC, M., 1994, Effect of a mixture of 15 commonly used pesticides on DNA levels of 8-hydroxy-2-deoxyguanosine and xenobiotic metabolizing enzymes in rat liver, *Journal of Environmental Pathology, Toxicology and Oncology*, 13, 163–68.

McELDOWNEY, S., HARDMAN, D.J. and WAITE, S., 1993, *Pollution, Ecology and Biotreatment*, Harlow: Addison Wesley Longman.

MONK, M., 1995, Epigenetic programming in differential gene expression in development and evolution, *Developmental Genetics*, 17, 188–96.

MORIMOTO, R.I., TISSIERES, A. and GEORGOPOLOS, C., 1990, *Stress Proteins in Biology and Medicine*, Cold Spring Harbor, NY: Cold Spring Harbor Laboratory Press.

MORIMOTO, R.I., TISSIERES, A. and GEORGOPOLOS, C., 1994, *The Biology of Heat Shock Proteins and Molecular Chaperones*, Cold Spring Harbor, NY: Cold Spring Harbor Laboratory Press.

NARAYAMA, P.K., GOODWIN, E.H. and LEHNART, B.E., 1997, α-particles initiate biological production of superoxide anions and hydrogen peroxide in human cells, *Cancer Research*, 57, 3963–71.

PHILLIPS, D.H. and VENITT, S., 1995, *Environmental Mutagenesis*, Oxford: Bios Scientific Publishers.

RUTHERFORD, S.L. and LINDQUIST, S., 1998, Hsp90 as a capacitor for morphological evolution, *Nature*, 396, 336–42.

SCANDALIOS, J., 1990, *Genomic Responses to Environmental Stress*, San Diego, CA: Academic Press.

SCANDALIOS, J., 1997, *Oxidative Stress and the Molecular Biology of Antioxidant Defenses*, Cold Spring Harbor, NY: Cold Spring Harbor Laboratory Press.

SHEN, H.-M., SHI, C.-Y., SHEN, Y. and ONG, C.-N., 1996, Detection of elevated reactive oxygen species level in cultured rat hepatocytes treated with aflatoxin B_1, *Free Radical Biology and Medicine*, 21, 139–46.

STEINGRABER, S., 1998, *Living Downstream*, London: Virago Press.

TERBOROUGH, J., 1992, *Diversity and the Tropical Rainforest*, New York: Scientific American Library.

WALDRON, K.W., JOHNSON, I.T. and FENWICK, G.R., 1993, *Food and Cancer Protection – Chemical and Biological Aspects*, Cambridge: Royal Society of Chemistry.

WELLBURN, A., 1994, *Air Pollution and Climate Change, the Biological Impact*, Harlow: Addison Wesley Longman.

WHITE, A., 1998, Children, Pesticides and Cancer, *The Ecologist*, 28, 100–105.

WILSON, E.O., 1992, *The Diversity of Life*, London: Penguin Books.

ZIMMERMAN, M., 1997, *Science, Nonscience, and Nonsense, Approaching Environmental Literacy*, Baltimore, MD: Johns Hopkins University Press.

Index